Incarnation and Neo-Darwinism

Evolution, Ontology and Divine Activity

— DAVID O. BROWN —

Sacristy Press

Sacristy Press
PO Box 612, Durham, DH1 9HT

www.sacristy.co.uk

First published in 2019 by Sacristy Press, Durham

Copyright © David O. Brown 2019
The moral rights of the author have been asserted

All rights reserved, no part of this publication may be reproduced or transmitted in any form or by any means, electronic, mechanical photocopying, documentary, film or in any other format without prior written permission of the publisher.

Scripture quotations, unless otherwise stated, are from the New Revised Standard Version Bible: Anglicized Edition, copyright © 1989, 1995 National Council of the Churches of Christ in the United States of America. Used by permission. All rights reserved worldwide.

Every reasonable effort has been made to trace the copyright holders of material reproduced in this book, but if any have been inadvertently overlooked the publisher would be glad to hear from them.

Sacristy Limited, registered in England & Wales, number 7565667

British Library Cataloguing-in-Publication Data
A catalogue record for the book is available from the British Library

ISBN 978-1-78959-060-9

This book is dedicated to the memory of Kenneth Peter Tully (1936–95) and Grace Eileen Brown (1926–2001).

Acknowledgements

This book started life fifteen years ago. Since beginning my undergraduate degree in Theology in Canterbury, not a day has passed in which I haven't spent time pondering those questions that my heroes spent their lives thinking and writing about: who is Jesus Christ, and what does he mean to our ever-changing world? This book represents the culmination of my own wrestling with those questions so far.

There are a number of people who have helped me through this journey that I would never be able to repay. I hope that my recognition of their help goes some way to showing them how grateful I am.

I would like to thank Dr Ralph Norman and Dr Jeremy Law, who helped and guided me through my doctoral studies, and Debra Wolter, who read through a draft of this book and offered comments and improvements.

I would also like to thank Richard Hilton, Natalie Watson, and all at Sacristy Press for having faith in my work and publishing it.

I would also like to thank Luke Underdown, the community at Worth Abbey, my wonderful partner Nuria Tain Novo, and my family, for their support while writing this book.

Lastly, I would like to thank my mum and dad. Without their constant, unconditional and unwavering support—both emotionally and financially—I would never have been able to achieve what I have.

David O. Brown
London, October 2019

Contents

Acknowledgements iv
Introduction 7

Chapter 1. The Neo-Darwinian Synthesis. 21
Chapter 2. Theological Anthropology. 47
Chapter 3. Divine Activity. 76
Chapter 4. The Person of Christ. 103
Chapter 5. The Incarnation. 120
Chapter 6. Participation, Imitation, and Neo-Darwinism 144
Chapter 7. The Cross 176
Chapter 8. The Resurrection. 198

Bibliography 234
Notes 256
Index 263

Introduction

It has been said by some, such as George Maloney, that "Christian theology is in crisis". The reason for this apparent crisis, according to Maloney:

> [Is] not because of its unchangeable doctrines, but mainly because the modes of representing the essentials of Christianity do not have meaning any longer for urban man. (Maloney, 1968, p. 224)

While it is not entirely clear that Christian theology *is* in crisis, George Maloney is absolutely correct in his evaluation of the situation. The perception that such a crisis exists does not indicate that theology is becoming increasingly incorrect—in the sense that science is gradually disproving its "unchangeable doctrines"—but rather that it is becoming increasingly incoherent to the modern world.

More often than not this incoherence is not due to a reluctance to engage with the modern world (although there are those, for example Young-Earth creationists, that do reject the theory of evolution in any and every guise). At best it is a misunderstanding of what modern science claims about the world, and, at worst, it is a refusal to take science at its word—a censoring of what biologists *actually* claim and setting up a "straw man" of what theologians *want* biologists to claim.

This book is an attempt to correct that failure. It takes neo-Darwinism (a theory of evolution through differential survival of genetic mutations)—arguably the most important scientific development in the past one hundred and fifty years (with the possible exception of quantum mechanics)—and explores its impact on theology. Such an exploration has been attempted before, but none of these attempts allowed neo-Darwinism an uncensored voice. Whether through misunderstanding or refusal to accept what neo-Darwinists actually say, very few, if indeed any, theological conversations with neo-Darwinism have engaged with an unbiased, uncensored, sober understanding of what neo-Darwinism actually concludes.

In having this conversation, this book will suggest ways in which theology needs to change emphases and nuances about what it says about the world, and, more specifically, what Christ's role in that world is; it will also explore ways in which neo-Darwinism can be seen to support the core message of theology, albeit using very different language. Christian theology can only be relevant to the modern world if it *engages* with it; this does not mean that theology must become unrecognizable, only that it must translate what it has to say into language that makes more sense to the modern world.

Theology and neo-Darwinism

It is almost certainly true that no idea has proved more of a burden to religion in the modern world than evolution. While the theory that species are mutable had been proposed as early as ancient Greece, it was not until Charles Darwin proposed his version of the idea that evolution became a dominant force. Of course, theology has faced important challenges before; however, what is unique about evolution is that most theologians do not consider it to be a problem. While there are clearly those who completely reject the idea of the mutability of species (normally as a result of taking a literal interpretation of the creation myth[s] in the opening chapters of Genesis), most theologians do not actually deny evolution and see evolution and theology as completely compatible.

Rather, it is scientists and atheist philosophers who see in evolution a proof against religious doctrines. As the English biologist St George Mivart writes, evolution "has been made use of as a weapon of offence by irreligious writers" and "thrown in the face of believers with sneers and contumely" (Mivart, 1871, p. 25). It was scientists who decided that evolution was incompatible with religion, not theologians. Other than the very few theologians who do not accept evolution, theology has never had much of a problem with evolution.

However, this overwhelming acceptance of evolution by theologians does not mean that they either understand it correctly or use a correct version of it in conversation with their theology. The detail of what neo-Darwinism actually claims, and thus the reason why theologians misunderstand or misuse evolution, is explored fully in Chapter 1. It can be said here, however, that almost without exception, theologians understand evolution in one of two ways: (a) evolution is directly controlled by God in order to create, or (b) God sets up the conditions of evolution and imbues it with certain values to indirectly create for God.

In other words, theologians understand evolution as being *teleological* (i.e. it is working towards a specific and achievable goal). By arguing for a directed, teleological understanding of evolution they contradict neo-Darwinism. Neo-Darwinism is not neutral on the subject of direction, as many claim: it is just that the evidence for this lack of direction is ignored or misunderstood by theologians. Further, by arguing that evolution is teleological (i.e. it will be finished at a specific point in the future) or as a paradigm from which salvation is needed, theologians conclude that evolution is a temporary phenomenon. Neo-Darwinism disagrees.

Perhaps the first writer to take on Darwin's newly published theory from a theological perspective was St George Mivart in his *On the Genesis of Species* (1871). Mivart's main bone of contention is that "natural selection is incompetent to account for the incipient stages of useful structures" (Mivart, 1871, p. 34). Natural selection, he claims, cannot by itself explain the wonderful diversity in the world. Evolution, then, "can have nothing to do with absolute or primary creation" and "is simply the divine action by and through natural laws" (Mivart, 1871, p. 269, pp. 278–9, p. 279). Darwinism, Mivart claims, is insufficient on its own and must be controlled by God in order to be truly creative.

Many theologians take this approach. George Barton, for example, in *Christ and Evolution* (1934), argues that one cannot accept the doctrine of evolution without believing that the "process of creation . . . possesses a purposive will" (Barton, 1934, p. 38). Quite simply, then, evolution is "God's way of doing his creative work" (Barton, 1934, p. 83). Phillip Hefner, too, in *The Human Factor* (1993), writes that "[evolutionary theory] is God's process of bringing into being a creature who represents the creation's zone of a new stage of freedom", which, he accepts, "amounts to interpreting the evolutionary process as the work of God" (Hefner, 1993, p. 32). Arthur Peacocke, a scientist-cum-Anglican priest, in *Paths from Science Towards God* (2001), also gives voice to the fact that the processes of the world "are to be seen as the very action of God" (Peacocke, 2001, p. 86). Cynthia Crysdale and Neil Ormerod also explain the outcome of evolution as being down to the influence of God. In *Creator God: Evolving World* (2013), they write that "with perfect intelligence God grasps all possible worlds ... God chooses one possibility in its totality from its beginning to its final consummation, from all myriad options" (Crysdale and Ormerod, 2013, p. 55).

Other theologians have also questioned the sufficiency of neo-Darwinism by endorsing a rival theory of evolution. Pierre Teilhard de Chardin, for example, saw Darwinism as being only part of the answer, understanding evolution to be a synthesis of Lamarckian and Darwinian evolution. Easily the most influential theologian to tackle the subject of evolution, Teilhard writes in *The Phenomenon of Man* (1955) that the chance that characterizes Darwinism is "recognized and grasped—that is to say, psychically selected" (Teilhard de Chardin, 1959, p. 149, n. 1). Again, then, for Teilhard, evolution is directed by God.

Ebenezer Griffith-Jones, in *The Ascent Through Christ* (1899) also expounds a theological interpretation of evolution that is remarkably similar to Teilhard. For Griffith-Jones, evolution is the "working of [God's] will" (Griffith-Jones, 1909, p. 36) and God's "method of working" (Griffith-Jones, 1909, p. 68), which represents an unmistakable ascent until it reaches humanity, whereupon the physical evolution becomes arrested—the human body representing the "climax" of evolution (Griffith-Jones, 1909, p. 50)—the "sway" of natural selection is "broken" and a "new principle of survival" comes into play (Griffith-Jones, 1909, p. 51).

Celia Deane-Drummond also explicitly turns to another theory of evolution. In *Christ and Evolution* (2009) she claims that her theory will "allow for the directionality of Conway Morris" and "accommodate the idea of punctuated evolution" (Deane-Drummond, 2009, pp. 22–3), both of which this book will argue misses the point of neo-Darwinism. John Polkinghorne, too, whilst not explicitly stating that God directly controls evolution, indicates a certain directionality to evolution when he asks, in *Science and Creation* (1988), that "every theory of evolution needs to account for what drives change forward" (Polkinghorne, 1988, p. 65). This book will show that not only does evolution not drive change forward, it actively *resists* any change that does occur.

This claim that evolution is driven forward, without explicitly stating that God has direct control, leads to another important facet of theological attempts to deal with evolution, namely that God sets up evolution to create for God. Dennis Edwards, for example, in *The God of Evolution* (1999), also questions the sufficiency of Darwinism. However, he does this not by questioning the completeness of the theory itself but by questioning the ability of biologists (and scientists in general) to detect the teleology he sees as inherent in evolution: Edwards did not doubt the sufficiency of Darwinism so much as question the scope of the scientific method altogether. He writes that it is "quite possible to think theologically of God as working purposefully in the universe through processes such as random mutation and natural selection, which when investigated empirically do not reveal purpose at all" (Edwards, 1999, p. 47). Edwards is explicit: the theologian is privy to something that the scientist could never detect. Of course, Edwards is correct that absence of evidence is not evidence of absence, but this book will demonstrate that there *is* evidence of absence, which most theologians ignore.

Fernando Canale, too, in *Creation, Evolution, and Theology* (2009), approaches the question of the relationship between science and theology by limiting the scope of the scientific method. Biologists, Canale writes, "have discovered only microevolutionary patterns that fall short of the macroevolutionary progress essential to evolutionary theory" (Canale, 2009, p. 72), making the bold claim that "no utterly convincing case of true speciation has yet emanated from a genetics lab" (Canale, 2009, p. 85).

In *Science and Religion* (2006), Holmes Rolston III writes that "science deals with causes, and religion deals with meanings" (Rolston, 2006, p. xiii). He also states that there is "a marvelous endowment of matter with a propensity towards life" that is "seen at a deeper level as the divine creativity" (Rolston, 2006, pp. 113–14). John Haught also subscribes to "[l]et[ting] science be science, and simultaneously [letting] theology be theology", arguing that whilst natural phenomena may have a natural explanation, this does not exclude "divine creativity as an ultimate explanation at a deeper level" (Haught, 2010a, pp. 22-4). It is not that Darwinism is incomplete and needs God to complement the theory, but that the scientist cannot detect a deeper teleology inherent within evolution.

There are also a number of theologians who disagree with the neo-Darwinian synthesis in that they hold it to be a temporary condition of the world from which creatures require salvation. Griffith-Jones, above, who saw humanity as the climax of evolution, is an example of this. Ted Peters and Martinez Hewlett, too, in their *Evolution From Creation to New Creation* (2003), see evolution as temporary, writing that "creation is still underway, not yet complete, not yet what God in Genesis would deem 'very good'" and that "creation requires redemption ... God has promised that death is going to be replaced by resurrection, and suffering will be no more" (Peters and Hewitt, 2003, p. 158).

Jack Mahoney went even further, explicitly claiming that salvation from evolution is the reason for the incarnation. In *Christianity in Evolution* (2011) he writes that the death of Jesus "is a remedying for us humans of the evolutionary fact of death that requires the disintegration of some organisms in order to provide for the emergence of others" (Mahoney, 2011, pp. 50-1). For Mahoney, evolution is simply a narrow replacement of the Genesis narrative and has no wider implications; this book will show this to be a mistake.

Christopher Southgate writes in *The Groaning of Creation* (2008) that he believes "in God's eventual healing of creation, and that humans have a part to play in that healing" (Southgate, 2008, p. 116), suggesting a similar understanding of the role of evolution to Mahoney's. Alejandro Garcia-Rivera's *The Garden of God* (2009) also understands evolution in this way, arguing that "death is not the final answer for us [because it is not] natural for the human creature" (Garcia-Rivera, 2009, p. 20). Once again, this idea that death is only a temporary condition, characteristic of evolution, and which must be redeemed, does not fit the neo-Darwinian paradigm.

Fact versus mechanism

One of the reasons theologians make the mistake of attributing a teleology to evolution is that there is confusion between the *fact* of evolution (i.e. that species are mutable) and the *mechanism* by which it happens (i.e. how species mutate). Although evolution has been proposed since the time of ancient Greece, not all theories of evolution are compatible, and they propose different *mechanisms*. Taking this into consideration, Stephen Jay Gould writes that "Darwin feared that people might confuse fact with mechanism, and cite the unresolved debate about natural selection as a denigration of his greatest achievement in establishing the fact of evolution" (Gould, 1982, p. xvii; see also Birx, 1991, p. 60).

It is for this reason that "Darwin initially resisted the word [evolution]" (Gould, 1997, p. 137); Darwin did not want his theory to be confused with other theories of evolution, which all saw speciation as being directed and idolizing progress. Darwin did not believe in progress and saw speciation as being neutral in this regard.

Darwin's fear was well-placed. There are plenty of commentators who confuse the *mechanism* of natural selection with the *fact* of evolution, and thereby reject the *fact* of evolution on the supposed basis that Darwin's *mechanism* of natural selection is incorrect. However, there is also a problem with those who confuse the *fact* of evolution with the *mechanism* of natural selection. There are plenty of commentators who claim to be espousing and supporting a neo-Darwinian theory of evolution on the basis that they accept the *fact* of evolution; they assume they are agreeing with the *mechanism* of neo-Darwinism because they accept the *fact* of evolution.

This is not always an explicit failure to follow the *mechanism* of neo-Darwinism; mostly it takes the form of a failure to grasp all of the implications of what Darwinism proposes. This becomes a problem because theologians attempt to provide a theory of how evolution and theology are compatible, yet they use a theory of evolution with which most scientists would disagree.

This is the very locus around which this book revolves: what happens to theology when neo-Darwinism is taken entirely at face value. This mainly takes the form of a rejection of teleology and asks what theology looks like when the teleological dimensions are removed. Almost all theological attempts to engage with neo-Darwinism do so on the basis that it can retain a teleological bent. More often than not this is in order to argue for evolution as a description of how God creates, as outlined above. This book will show that this inclusion of teleology is a major and fatal distortion of Darwinism, which *cannot* accommodate a teleological bent; in fact, it actively argues against it.

Moreover, in this way most theological conversations focus on the "superficial" aspect of neo-Darwinism—the fact of species changing from one form into another and how that is compatible with the opening chapters of Genesis. Evolution is considered to be almost exclusively a doctrine of creation that has nothing more to contribute beyond that narrow context. Evolution is treated as a prologue, detailing how things came to be, before the real business of theology is attempted. This book will take a different approach. It will argue that neo-Darwinism is nothing less than ontology—it describes the very nature of being. This, it will be argued, is not just compatible with theology on the superficial level of describing how things came *into* being, but how things *are* generally.

Why neo-Darwinism?

All this being said, this book is not an apology for neo-Darwinism; it is not important *why* Darwinism should be favoured over other theories of evolution, only that it *is* favoured. In the chapters that follow, neo-Darwinism is used as a conversation partner for theology, because it is the theory of evolution that has enjoyed almost unanimously consistent support. As Friedel Weinert writes in *Copernicus, Darwin, & Freud*:

> On the threshold of the twenty-first century, many of Darwin's original difficulties have been cleared up. The theory has also increased in scope to explain such diverse phenomena as sexual and asexual reproduction, the sex ratio problem, and the evolutionary advantage of altruism. Evolutionary psychology claims an even greater scope for the theory, for it wants to explain

mental facts by reference to evolutionary principles. (Weinert, 2009, p. 173)

Neo-Darwinism is used in this way because it is the only theory of evolution that has consistently been supported and furthered by continuing scientific experimentation. *How* that support has been discovered is unimportant; the *fact* that it has that support is.

Moreover, so great is this continuing support and development that neo-Darwinism is the only theory of evolution that has been applied outside of the narrow biological context in which it was first proposed. The scope of neo-Darwinism has increased to such an extent that Richard Dawkins even proposed that Darwinism could be used to describe the "competition" between ideas, which he called *memes*. Thus, he writes that "genetic natural selection", which was identified by neo-Darwinism, is "only a special case of a more general process that I came to dub 'Universal Darwinism'" (Dawkins, 2004, p. 149). This adds further evidence to the idea above that neo-Darwinism can be used as a philosophical, and more specifically an ontological, position, rather than a narrow replacement of the opening chapters of Genesis. No other theory of evolution has maintained the consistent support that neo-Darwinism has and, as a result, no theory of evolution has had any success outside of the narrow biological setting in which it was first proposed. It is for this reason that neo-Darwinism is used, and why other theologians are criticized for not following a strict neo-Darwinian approach.

The method and scope of the book

The objective of this book is to explore what happens to theology when neo-Darwinism is taken seriously and exactly as biologists expound it. More specifically, it will explore the role of Christ in a neo-Darwinian, theological paradigm. Christ is of particular importance to this question because (a) the rejection of a literal interpretation of the Genesis narrative (i.e. the Fall of humanity) appears to reject the very need for Christ in the first place (i.e. salvation), and (b) Christ will be shown to be the solution to the problem of how God interacts with and influences creation in a neo-Darwinian paradigm that rejects direct divine involvement. This book will answer the question of what Christology looks like when neo-Darwinism is allowed to set the boundaries and the framework of theological discourse.

Although this book is primarily an exercise in systematic theology, it also falls into the category of "religion and science", a theological discipline that has risen in prominence in recent years. Any theological question that arises within this remit will necessarily be quite broad, as such a large subject area cannot possibly be dealt with in a single volume. There are so many aspects to theology and, likewise, so many different scientific disciplines within which one could attempt to answer any theological question. This book, therefore, will narrow the parameters of its discussion to the essential issues.

a. Science

From the scientific perspective, the scope of this book is limited solely to evolution and, more specifically, neo-Darwinism. There are a number of reasons for concentrating on the contribution of evolution. Perhaps the most important is that many scientists and philosophers have claimed it is the final coffin nail for theology. Evolution, according to them, closes the last gap in which God could hide.

Evolution has become very much part of the wider social consciousness; due to the success of scientists such as Richard Dawkins, the non-scientific public assumes that evolution creates problems for theology. Conversing with neo-Darwinism, therefore, becomes almost an exercise in apologetics; it is important to show that theology can contribute to the conversation regarding theology's supposed demise.

Neo-Darwinism is also the sole scientific focus of this book because it is the theory of evolution that has enjoyed the most success and evidential support: the scientific community has continually put it to the test, and it has consistently made the grade, gradually being further refined. However, neo-Darwinism is also important as a conversation partner for theology, because it is a paradigm that has outgrown the narrow biological field in which it was first proposed as a solution to the problem of the mutability of species, and has become a philosophical position in its own right.

For these reasons, this book, whilst conversing with a scientific theory, will not employ a scientific *method* in expounding a theology. As noted earlier, it is not concerned with *why* neo-Darwinism has triumphed over other theories of evolution and has come to enjoy pride of place within today's scientific community, but only that it *has*. In this way, this book is more concerned with the *philosophical* implications of neo-Darwinism than with the *scientific*.

Neo-Darwinism, then, occupies the place in modern theological discourse that Platonism and Aristotelianism held in Patristic and Scholastic theology respectively. As Ilia Delio asks, quoting Zachery Hayes, if Augustine spoke theologically "in a world conditioned by Neo-Platonism", and Aquinas spoke theologically "using Aristotelian categories", then "is it possible for contemporary theology to do a similar thing, taking a world view from the sciences?" (Delio, 2013, p. xx). Neo-Darwinism simply represents the new language in which theology will be discussed.

Although the scientific scope of this book is limited to neo-Darwinism, other important scientific discoveries will also be appealed to; for example, in Chapter 3, "Divine Activity", the idea of relativity and the related non-separateness of time and space will support the conclusion that is reached. However, further scientific disciplines and ideas are used *only to support* the original point that is raised by neo-Darwinism. In this particular instance, the use of modern physics, primarily those ideas initiated by Einstein, adds support and further evidence to the notion that God does not, and cannot, influence the direction of evolution.

b. Theology

The theological side of the relationship is much more important for this book. This is, after all, a work of theology that is informed by neo-Darwinism, not a work of neo-Darwinism that is informed by theology. It is precisely because the relationship is understood in this manner that neo-Darwinism will be accepted without any theological influence. This means that neo-Darwinism will be taken as it is expounded by biologists and scientists, rather than the sort of neo-Darwinism that has already been criticized, namely, neo-Darwinism that has been passed through a theological lens and all its "unfavourable" elements filtered out.

Moreover, this is a work of Christology, which means that it is primarily concerned with the role and contribution of Christ in this neo-Darwinian paradigm. The systematic nature of theology means that it is difficult to consider the role of Christ without reference to other facets of theology—such as the Church, the Holy Spirit, and the sacraments. However, these elements of theological endeavour are not dealt with in this book; instead, its scope is limited to the question of who Christ is and what relationship he has with creation, and how those questions are answered and made meaningful in a neo-Darwinian paradigm.

As has already been noted, the reasons for concentrating on the role of Christ are (a) in order to address the rejection of the paradigm of salvation that evolution seems to demand, and (b) because Christ becomes the solution to the problem of divine activity in a neo-Darwinian paradigm. However, there are other important reasons. Despite the eventual Trinitarian nature of Christian doctrine, it is the belief in the divinity of Christ that ultimately separates Christianity from the other Abrahamic religions. In this respect, it is this belief that is most distinctive about Christianity. How neo-Darwinism influences Christianity, therefore, must surely start with how it influences the core doctrines and beliefs about Christ.

The doctrine of the Trinity further reinforces the concentration on the role of Christ in another important direction. The Great Schism between the East and West in 1054, whilst concerned with a multitude of practical and political issues, is best remembered for the argument over the place of the *filioque* in the Creed. The East rejected the *filioque*, whereas the West retained it. In the West, this gave Christ, despite protestations, a "grander" place in Trinitarian relationships. This has led to Western theology, or at least certain manifestations of it, being labelled "Christocentric" (Boff, 1997, pp. 166–7). Essentially, this means that the role of the Spirit is only ever conceived within the context of the role of the Son and only ever conceived as an extension or continuation of the mission of the Son.

However, perhaps more pertinently, this book will argue that the neo-Darwinian synthesis itself dictates that Christocentrism—i.e. the central and primary role of Christ—is the correct path to follow. The central place of Christ, therefore, may well be dictated by other concerns, but it is also strengthened and supported by neo-Darwinism.

The shape and argument of the book

Part I begins with an exposition of neo-Darwinism as it is understood by biologists. The intention of Chapter 1 is to show that, despite the claims of many theologians, neo-Darwinism does not remain neutral on the subject of teleology. It will consider the evidence for an interpretation of neo-Darwinism that actively resists change, leading to the conclusion that evolution cannot be a goal-oriented and/or directed phenomenon. It will also argue that neo-Darwinism, because it is not culminating in a future goal, is not a temporary phenomenon that will be completed at some point in the future, but a permanent condition of what it means to be created.

Chapters 2 and 3 deal with the implications that this interpretation of neo-Darwinism has for anthropology (i.e. doctrines of humanity) and theology (i.e. doctrines of God), specifically divine activity. From the perspective of anthropology, neo-Darwinism denies the uniqueness of humanity in terms of its relationship to the rest of creation and in terms of its relationship with God. Humanity, from a neo-Darwinian perspective, cannot be understood as an improvement on the rest of creation, and thus ontologically superior. Rather, all creatures are considered equal. Furthermore, humanity can no longer be considered different in its relationship with God. The Fall of humanity must be rejected and the special or unique need for grace along with it. This, however, does not lead to a rejection of the doctrine of original sin, which is not only retained but is also seen to align and agree with the conclusions of neo-Darwinism.

From the perspective of divine activity, neo-Darwinism rejects the assertion that God can direct evolution specifically and the universe generally. Not only is this supported by philosophy and modern physics, but it also coheres with traditional and classical understandings of the transcendent nature of God. God, utterly transcendent and utterly different, i.e. immaterial and eternal, cannot directly influence the universe. If God wants to influence the world, God needs to be created; Christianity already has a doctrine that claims as much, namely, the incarnation.

Part II of the book deals with the role of Christ. Chapter 4 considers the hypostatic union and, concentrating on the role of *communicatio idiomatum,* further expounds how Christ can be considered the unique agent of divine influence on the world. This leads to a consideration of the incarnation, in Chapter 5, as *the* divine act of creation. In conversation with the Jewish doctrine of *tzimtzum*, it will be shown that the incarnation can be seen as the eternal divine activity that gives being to the universe and explains the openness to failure that characterizes "universal Darwinism", thus accounting for how the universe can spontaneously "pop" into existence from nothing, accidentally.

Chapter 6 continues this discussion of Christ as the unique agent of divine activity by considering the role of participation. It will be shown that the doctrine of participation is not only synonymous with that of imitation of Christ, but also with the ontology that characterizes neo-Darwinism. From this point of view, it will be demonstrated that neo-Darwinism fundamentally agrees with what Christian theology understands to be ontology.

No discussion of Christology is complete without reference to the cross and the resurrection, and so it is important to explore what neo-Darwinism means for them. In Part III, Chapter 7 considers the cross and shows that it can satisfactorily be removed from the paradigm of salvation and redemption in which it has traditionally been cast. In this way, this chapter helps to inform and further nuance the chapter on the incarnation (Chapter 5) and the idea of divine activity through self-sacrifice and *kenosis*.

This leads to a discussion of the resurrection, in Chapter 8, which, it will be argued, is the biggest stumbling block to a theology that is sensitive to neo-Darwinism. Rather than understand resurrection as the future raising of a perfect and incorruptible body, neo-Darwinism must re-cast what the resurrection means. Drawing on the fact that the resurrection must be primarily concerned with relationship with God rather than prolongation of human life, the resurrection can be seen within the scope of participation with and imitation of Christ.

CHAPTER 1

The Neo-Darwinian Synthesis

It has long been beyond reasonable doubt that evolution, as a general theory of speciation, is correct. However, the particulars of that general theory have been the subject of much discussion over the years, especially following the publication in 1859 of Charles Darwin's *On the Origin of Species*. Despite this ongoing debate, it is more or less universally agreed that neo-Darwinism is the best theory to explain *how* evolution proceeds. However, the *fact* of evolution (i.e. *that* it happens) is often confused with the *mechanism* of evolution (i.e. *how* it happens) and this leads to misconceptions.

Stephen Jay Gould notes that when Darwin wrote *Origin* he was concerned that people would confuse the *fact* of evolution with his proposed *mechanism* of natural selection (Gould, 1982, p. xvii; see also Birx, 1991, p. 60). Darwin's concern was entirely justified. Even in the twenty-first century, one hundred and fifty years after the publication of *Origin*, the *fact* of evolution is considered synonymous with the *mechanism* of natural selection and is judged entirely on that basis. Thus, creationists (and other doubters of evolution) claim that evolution is incorrect, based on the assumption that natural selection is insufficient to account for the diversity and apparent design in the world.[1]

However, there is a more pressing problem with the conflation of the *fact* of evolution and the *mechanism* of natural selection, namely, that those who accept evolution are assumed to be conforming to the neo-Darwinian synthesis by default. This creates a situation whereby theological interpretations of evolution are assumed to be supported by the science simply because they claim to accept the validity of evolution, when, in fact, they make claims that contradict neo-Darwinian biology.

Acceptance of the *fact* of evolution does not guarantee adherence to the *mechanism* of Darwinism.

One of the most important assumptions made by theologians (which this chapter will dispel) is that Darwinism can be interpreted as displaying a direction or teleology (or that certain values are cultivated). However, as Gould notes:

> *Evolution* entered our language as the favoured word for what Darwin had called "descent with modification" because most Victorian thinkers equated such biological change with progress ... Darwin initially resisted the word [evolution] because his theory embodied no notion of general advance as a predictable consequence of any mechanism of change. (Gould, 1997, p. 137)

Those who accept evolution as correct do not automatically conform to the neo-Darwinian synthesis. Those who argue that their theological interpretations of evolution conform to neo-Darwinism, yet still see evolution as directed, are incorrect.

It is unimportant for the purposes of this chapter to answer the question of *why* neo-Darwinism is the favoured theory of evolution; it is simply enough to note that there is increasing support for it, with laboratory studies producing sufficient "direct observation" of natural selection (see Peacocke, 2001, p. 66 and Bynum, 2009, pp. xlix ff.), and that many of the problems with it have been cleared up (Weinert, 2009, p. 173). Moreover, there is little, if any, evidence for other theories, for example, Lamarckism (Elliot, 2012, p. 10 and Huxley, 1942, pp. 457ff.).

The neo-Darwinian synthesis is understood as being the theory of natural selection combined with the modern theory of genetics (O'Leary, 2007, p. 132). Genetics provides the answer to Darwin's ignorance of "whatever the cause may be of each slight difference in the offspring from their parents" (Darwin, 2009, p. 157). Genetics, then, completes the unfinished elements of Darwin's theory. The discovery of the gene also led to the identification of the important difference between the genotype and the phenotype. Put simply, the phenotype is the manifestation or expression of the particular "thing" that the gene (i.e. genotype) codes for (Dawkins, 1999, p. 83). In this way, blue eyes are the phenotype of which the particular DNA code that expresses them is the genotype.

In simple terms, Larmarckism is the theory that evolution is due to changes in the phenotype (which can be moulded by the environment),[2] and Darwinism is the theory that evolution is due to changes in the genotype (Dawkins, 1999, p. 167; McGrath, 2005, p. 127). This is an important distinction to make. Neo-Darwinism, as will be shown in this chapter, is the blind selection of genotypes based on the survival of phenotypic expressions, yet these phenotypic expressions cannot influence mutations and variations in the genotype.

This chapter will provide an unbiased and uncensored outline of what biologists consider the neo-Darwinian synthesis to be. The core of this will be to show that evolution is not directed or controlled by God, Moreover, it will show that neo-Darwinism is far from being neutral about the role that God plays in evolution—i.e. that Darwinism describes *how* and not *why*—and will provide direct evidence *against* this position. Firstly, the development of neo-Darwinism will be briefly considered with a presentation of the major discoveries, particularly those that have implications for the role of God. This will be followed by some modern criticisms of Darwinism. By highlighting other ways of presenting evolution, those important distinctions that set Darwinism apart will be emphasized. Lastly, the theological and philosophical implications of the neo-Darwinian synthesis will be considered.

The development of neo-Darwinism

Much could be written about the many biologists who have contributed to the development of neo-Darwinism; however, there are certain developments that highlight how biologists understand neo-Darwinism and how this differs from the neo-Darwinism that theologians assume they are following.

Along with the rejection of teleology, or the affirmation of teleological neutrality of evolution (see Deane-Drummond, 2009, p. 229), the most important element of this understanding is that evolution is not a theory of *change*; instead, it is a theory of *preservation*. Indeed, Darwin himself was explicit in this regard, writing that

> [T]his principle of preservation, I have called, for the sake of brevity, natural selection. (Darwin, 2009, p. 121; see also Haught, 2010a, p. 32)

While neo-Darwinism was a theory that attempted to explain the reason for change and diversity in the world, it actually explains that this change is an accident that is caused by a mistake in replication; it is only when genes fail at staying the same that change occurs. Change is always secondary to preservation and is in fact explained by it; without the primacy of preservation, evolution could not occur. Focusing on this primacy of preservation, this chapter will show how developments in neo-Darwinism provided increasing evidence for understanding all change as accidents in replication, and how this rejects the notion that evolution is in any way guided.

———

The history of neo-Darwinism obviously starts with Darwin himself. While he did not invent the theory of evolution, he did present the idea of natural selection as its mechanism. Again, much could be written about Darwin's great contribution, but three important elements will be considered here that demonstrate the primacy of preservation, and thus where most theologians misinterpret evolution. These are (a) the subjective nature of variation (i.e. the absence of "correct variations"), (b) the non-creativeness of natural selection (i.e. natural selection does not cause mutation), and (c) the accumulative nature of these variations (i.e. the absence of "intermediate individuals").

a. The subjective nature of variation

The subjective nature of variation means that natural selection does not have a set of criteria against which it judges every individual in order to ensure progress or anything else. Darwin writes that

> [N]aturalists have not yet defined to each other's satisfaction what is meant by high and low forms. (Darwin, 2009, p. 297)

Evolution does not produce better and better organisms: it simply happens that due to there being more individuals than the resources can support (a principle that Darwin learned from Thomas Malthus' observation that populations grow geometrically, whereas resources grow arithmetically [see Birx, 1991, pp. 59–60, p. 135]), some organisms can survive longer, and reproduce more, than others. Natural selection, then, does not aim to produce perfect beings—it simply notes that, since death and extinction are inevitable (because resources are limited), those individuals that do survive have more offspring, and so thrive.

This point can be perfectly illustrated by a mistake in interpretation. Alan Lacey asks:

> Natural selection may eliminate unwanted variations ... [but] what guarantees that the right variations arise? (Lacey, 1989, p. 180)

Not only is natural selection completely incapable of guaranteeing anything, there is also no such thing as "unwanted variations" or "right variation". The "correct" variation is simply the one that is able to survive long enough to reproduce. Thus, as Ronald Cole-Turner argues, "when the environment changes, its selection criteria will change" (Cole-Turner, 1993, p. 43). Or, better (since selection is not a real phenomenon but simply an observation of which individuals survive), when the environment changes, other individuals will survive and reproduce.

Natural selection is not an active process that seeks the best solutions to particular problems; it is simply the observation of differential survival (Dawkins, 1998, p. 18)—or, perhaps better, "differential reproduction" (Monod, 1972, p. 115)—and if one variation is able to "tip the balance" of survival then this will survive longer and reproduce more than other variations. Crysdale and Ormerod illustrate this perfectly:

> Nature does not select anything ... [instead] it ought to be considered a process of natural elimination. (Crysdale and Ormerod, 2013, p. 36)

b. The non-creativeness of natural selection

Natural selection is simply, and *only*, the observation that some variations are better able to survive and reproduce; it is *not* a directed or value-imbued process that helps to ensure the correct variations arise. This recognition that natural selection is only an observation of differential survival calls attention to another important element of Darwin's theory that is often overlooked: the non-creativeness of natural selection. What this means is that the environment in which individuals find themselves cannot provoke variation, let alone a particular variation. Therefore, natural selection can only modify what it already has—it cannot create anything new. Instead, natural selection has to wait for a variation to accidentally emerge (through genetic mutation, which will be explored below), which it can then mould. This leads Patricia Williams to say:

> Evolution is a tinkerer and has never designed organisms completely from scratch (even in the beginning, it had to use available chemicals and their possible interactions). (Williams, 2001, p. 137)

Again, natural selection only preserves (or eliminates)—it does not create anything. This principle, subsequently confirmed experimentally by August Weisman (who showed that the environment [i.e. selection criteria] cannot cause genetic mutation [see Weisman, 1893, pp. 392–4]), is the very same point that was made previously: that the phenotype is incapable of influencing the genotype (thus, it provides laboratory evidence for the incorrectness of Lamarckism).

The non-creativeness of natural selection is further illustrated by what can be termed "form before function".[3] What this means is that anything that occurs as the result of evolutionary change (for example, an organ) arises first—*before* it is put to use by the individual. Or, to put it differently, the "function is the *effect*, not the *cause*, of an organ" (Weinert, 2009, p. 162, italics added). The environment cannot provoke a variation to occur, let alone a particular or correct variation, because of the need for a particular function. Darwin himself noted this exact idea, writing:

> The illustration of the swim bladder in the fishes is a good one, because it shows us clearly the highly important fact that an organ originally constructed for one purpose, namely flotation, may be converted into one for a wholly different purpose, namely respiration. (Darwin, 2009, p. 174)

When life emerged from the oceans, it did not take note of its new environment and then evolve a respiratory system in order to cope with it. Rather (to put it somewhat simplistically), a new variation occurred that enabled the swim bladder to survive outside the oceans, and so life was able to move into a different environment. The variation happened first, by accident, and was only then "selected" for a different use.[4]

c. The accumulative nature of variations

The third element, one that is often overlooked by theologians when discussing evolution and neo-Darwinism, is arguably the most important: the role of accumulation. Darwin writes that "*natura non facit saltum*" (Darwin, 2009, pp. 177–8), or "nature makes no leaps". Evolutionary change happens over an extended period of time due to the *accumulation* of small variations, or "little by little" (Bynam, 2009, p. xxxvii). This points to the primacy of preservation. Tiny variations are gradually accumulated until, eventually (and only noticeable due to the extinction of others [see Darwin, 2009, p. 165]), observable change occurs.

The role of accumulation is often denied on the basis that theologians disagree as to whether micro-evolution is the same as macro-evolution. Micro-evolution (i.e. apparently negligible or arbitrary variations), these theologians argue, is a different phenomenon to macro-evolution (i.e. major and observable species-defining variations): the small variations between different individuals of a species are not the same as the large variations between species. However, most biologists disagree with this claim.

Berry, for example, writes that "the evidence for infra-specific adjustment (micro-evolution) is overwhelming" (Berry, 1982, p. 47).

Major observable change is simply the accumulation of tiny variations that, by themselves, appear to be arbitrary (Collins, 2007, p. 132). Macro-evolution is only understandable through the comprehension of micro-evolution (see Dobzhansky, 1982, p. 12). This clearly points to the primacy of preservation; natural selection is constantly preserving what is already existent, accidently preserving the tiny variations (caused by mistakes in preservation).

This element can also be perfectly illustrated by a mistake. Smith writes that

> [W]hat is still more damaging to the evolutionist, however, is that under Darwinist auspices ... even in our age there should exist transitional forms, living species, therefore, which exhibit structures of a nascent kind. (Smith, 1988, p. 7)

In fact, *every* individual organism is a transitional form. Or, perhaps more precisely, no individual organism is a transitional form since evolution is not about transition. There is nothing towards which evolution is headed—there is no goal or *telos*, so there cannot be *transitional* forms, as if there are some imperfect organisms that are making the journey towards something better. Or, put differently: every individual is transitional in the sense that variation never stops.

Taken together, these three points (the subjective nature of variation, the non-creativeness of natural selection, and the accumulative nature of variation) lead to one conclusion about Darwinism that is not often appreciated by theologians and, more importantly, argues against their own interpretation. Evolution is not a process by which progress happens. Natural selection does not cultivate certain values, nor does it provoke certain "correct" variations to arise. What it does do is the complete opposite of this claim: natural selection *preserves*. If a variation does occur and is able to provide a slight differentiation in survival, then this is preserved and, eventually, enough slight variations are accumulated to produce change. However, this change is only the result of a failure to preserve perfectly. Change is not the intention—it is only a side effect of the imperfection of preservation.

For Charles Darwin, therefore, evolutionary change happens through the accumulation of minute and unintended variations. These variations do not represent progress; natural selection is not a teleological process. Likewise, there is nothing that can guarantee that any such variation will arise—there is no such thing as a correct mutation. All that natural selection can "do" (in fact, natural selection does not "do" anything—it is simply an observation) is eliminate those that are unfit and preserve those that are fit, and that is because more individuals are born than natural resources can support, thus creating a situation of differential survival. Any variation that gives an advantage will be preserved, and any variation that does not will be eliminated. Furthermore, what is considered an advantage is completely subjective, is 'qualitatively different for every different organism' (Fisher, 1930, p. 37), and will change as environments change (Cole-Turner, 1993, p. 43).

These ideas can be understood as representative of the core of Darwinism. Regardless of how much selection of mutations is appealed to as a theory of evolution, without the emphasis on the accidental and blind nature of this (i.e. the non-creative nature of it), this is not Darwinism. Darwinism is about preservation.

———

Theodosius Dobzhansky is another important contributor to neo-Darwinism. In his book *Genetics and the Origin of Species* he confirms that gene mutations and structural changes in chromosomes are the sources of the variations from which nature "selects" (Dobzhansky, 1982, p. 118). However, what is more important than this discovery is that the source of these mutations (which cause variation) is exactly the same as that with which biologists are already familiar (Williams, 2001, p. 73). Thus, Dobzhansky writes:

> Some critics have hastened to remark that since mutations and chromosomal changes can be induced by as destructive an agent as x-rays, such changes bring about degeneration and not evolution. The logic of this criticism is certainly rather ludicrous. (Dobzhansky, 1982, pp. 82–3)

It is difficult to overemphasize the importance of this claim. Those "destructive" effects on genes that are brought about with x-rays are precisely the same effects that are responsible for the genetic variation that is the raw material of natural selection.[5] The variation that occurs naturally (whose phenotypic expressions are subsequently selected) is *exactly* the same as those that can be provoked "artificially" in a laboratory by the deleterious effects of x-rays causing the genes to "malfunction". All mutations, which are the "principal source of variation" (Dobzhansky, 1982, p. 118), are due to a malfunction or failure of the gene to replicate itself properly.

This does not mean that such variation has a negative *effect*—Dobzhansky agrees that "classification of mutations into favourable and harmful ones is meaningless" (Dobzhansky, 1982, p. 23)—it has already been established that the judgement regarding the value of a variation is always subjective. Rather, it means that the *cause* of variation is always due to a mistake in replication. This mistake may lead to a favourable variation, but it is always brought about due to a failure to replicate properly. Speciation (i.e. observable evolutionary change) is simply the accumulation of these variations due to their differential survival. Once again, therefore, evolution is more concerned with preservation than with change. Change only occurs because the "process" of replication is not perfect and is open to failure.

―

Perhaps no biologist is more explicit in his or her support for the interpretation of neo-Darwinism that is offered in this book than Jacques Monod. In his immensely important and influential *Chance and Necessity*, he writes that "invariance necessarily precedes teleonomy" (Monod, 1972, p. 32). What this means, in simple terms, is that preservation must be present for any change or variation to have an effect. Without preservation, accumulation cannot happen. Without preservation, whatever variation does occur cannot hang around long enough to have any effect on the organism, positive or negative. As Monod says:

> The Darwinian idea [is] that the initial appearance, evolution, and continuous refining of ever more intensely teleonomic structures are due to disturbances occurring in a structure *which already possesses the property of invariance*—and hence is capable of preserving the effects of chance and thereby submitting them to the play of natural selection. (Monod, 1972, p. 32)

Again, he is emphatic: if genes do not *already* possess the ability to preserve themselves in replication, then no mutation, however advantageous or disadvantageous, will be present long enough to influence evolutionary change. For Monod, evolution is "random chance caught on the wing, preserved, reproduced by the blind machinery of invariance" (Monod, 1972, p. 96) and therefore when genetic mutations occur, "such errors of transcription, thanks to the blind fidelity of the mechanism, will be automatically reproduced" (Monod, 1972, p. 109).

Invariance, or preservation, is so powerful that any variation (i.e. error of transcription) is caught up in the relentless replication and preserved (i.e. accumulation). Moreover, since those "errors" are accidental, and "constitute the only possible source of modifications", then "chance alone" is the source of all creation (Monod, 1972, p. 110).

Moreover, according to Monod, genes actually resist such mutation:

> It might seem then that by virtue of its very structure, this system [of replication through DNA] ought to resist change, all evolution. This it certainly does . . . [therefore] living beings, despite the perfection of the machinery that guarantees the faithfulness of translation, are not exempt from this law [of physics that no microscopic entity can fail to undergo quantum perturbations]. (Monod, 1972, pp. 108-9)

This is very important. Not only do mutations occur accidentally—as the result of mistakes in copying—they are also actively resisted and happen in spite of efforts to prevent change. Therefore, evolution is not a directed process of change, and genes actually try to prevent change from happening. Monod, then, is absolutely clear: all variation is accidental, is the sole source of the raw material of natural selection and is only possible due to the primacy and high-fidelity of preservation and invariance.

Richard Dawkins, perhaps the most famous biologist of the early twenty-first century (although this is due more to his outspoken and vehement criticism of religion than for his biology), also continues to espouse this very interpretation of neo-Darwinism; evolution is simply "the struggle of gene lineages to replicate" (McGrath, 2005, p. 36). Dawkins agrees with Monod that "most of natural selection is concerned with preventing evolutionary change rather than with driving it" (Dawkins, 1986, p. 125). Change may occur, but "most new genes that arise, either by mutation or re-assortment or immigration, are quickly penalized by natural selection" (Dawkins, 2006, p. 86). Sometimes a variation can "tip the balance" just enough that it is able to reproduce and accumulate, and then, eventually, observable evolutionary change occurs. However, the presence of variation in the first place is only due to the fact that the "process" of replication is imperfect. Dawkins writes:

> No copying process is infallible ... the mutation brings into existence a new kind of replicator which "breeds true" until there is a further mutation. (Dawkins, 1999, p. 85)

This "new" gene, which has been blindly selected, is then blindly re-copied until such time as another mutation occurs. Regardless of how much genes preserve themselves and resist change, such change is inevitable at some point (yet, this inevitability does not guarantee that a particular mutation will occur).

Perhaps Dawkins' most famous, and most often denigrated, contribution comes from the title of one of his books, rather than the content of his arguments. The idea that genes can be *selfish* has often been criticized as ascribing anthropomorphic categories onto genes (see McGrath, 2005, p. 41). However, this completely misses the point. Dawkins does not mean that genes are selfish as opposed to being altruistic, but that they are selfish in the sense they are concerned with nothing other than replicating as faithfully as possible without allowing other variations to arise; or to put it differently: that genes are selfish in the sense that "they readily reproduce copies of themselves" (Williams, 2001, p. 130). Thus, Dawkins writes that a successful unit of natural selection must have 'longevity, fecundity, and copying-fidelity (Dawkins, 2006, p. 24).

If genes are not successful at copying themselves faithfully and plentifully, then they will not survive. In this way, Dawkins supports the theme of this chapter: evolution is primarily about preservation, *not* change; change only occurs when preservation fails. That change, the sole source of what is selected, is caused solely by the failure of genes to replicate properly and is both random and accidental (i.e. not caused by selection).

Non-Darwinian theories

There are those who disagree with the sole sufficiency of *blind* natural selection of *random* and *accidental* variations caused by genetic mutations to explain all the diversity and "apparent" design in the world. Considering their ideas serves to further illustrate the importance of this interpretation of neo-Darwinism.

a. Modern intelligent design theory

Perhaps the most famous of these disagreements is modern intelligent design (ID) theory. The most popular supporter of this position is Michael Behe, who argues that natural selection cannot possibly be powerful enough to account for the vast complexity of biological systems. Behe calls this "irreducible complexity", by which he means that some systems or organs are so complex—reliant on the presence of a number of different and interdependent components—that they could not possibly have come into existence slowly through gradual and random accumulation.

Behe uses the example of the mousetrap to explain his reasoning. He claims that if any part of the mousetrap were just a little bit shorter, it would be completely ineffective as a mousetrap and so would not be selected

(Behe, 1998, pp. 176–7). The mousetrap would be irreducibly complex and would have to emerge all at once as a distinct unit, or evolution would have to be guided, selecting the "intermediary individuals" on the basis of their potential and future suitability—two suggestions that are abhorrent to the neo-Darwinian synthesis. The same can be said, claims Behe, for such biological systems as the bacterial flagellum and the blood-clotting cascade, both of which represent irreducible complexity.

Essentially, Behe makes two mistakes in his criticism of neo-Darwinism. The first has already been encountered as "form over function". The particular components that make up any "irreducibly complex" system could have been employed in the past for different functions, or could be the by-products of other evolutionary favourable variations. The blood-clotting cascade, for example, can only be irreducibly complex if it was designed *only* to perform such a task. Neo-Darwinism denies that this can be the case—gene mutation and natural selection are not teleological—and the presence of the cascade can only be explained by the organism putting to use vestiges of previous forms.

The second mistake Behe makes is that he does not distinguish between genotypic and phenotypic replication. The phenotype (the mousetrap, for example) could have "leapt" into existence with only one of two genetic mutations. The phenomenon of polydactyly provides an example of this idea. A hand that contains a sixth finger did not have to "grow" the finger over numerous generations from a tiny stub to a full finger; the extra finger could have "sprung" up in one generation due to the mutation in one gene (see Miller, 1999, pp. 103ff.).

In other words, under neo-Darwinism, there may not need to be as many generations and mutations in order to produce the apparently designed organic structure (or mousetrap). This argument becomes more acute when it is combined with the above criticism regarding "form before function"; the generational distance between a form performing one function to its modification in performing another function may not necessarily be that great (see Berry, 1982, pp. 8–9).

As a result of these criticisms (and coupled with an assurance that God is solely responsible for removing irreducible complexity), it is often claimed that intelligent design is nothing more than "stealth creationism" (Collins, 2007, p. 183) with "some concessions to modern science" (Dupré,

2009, p. 169). Ronald Numbers even suggests that creationists simply did nothing more than replace the word "creation" with "intelligent design" in order to make their views seem more presentable (Numbers, 2010, p. 137). (Interestingly, Weinert also notes that there are some similarities between Lamarckism and creationism: for example, the presence of goal-oriented evolution or teleology [Weinert, 2009, p. 124].)

In other words, there is an instance here of the tension between *fact* and *mechanism* that was outlined above. The *fact* of evolution is no longer contested (except by a small minority of Young-Earth biblical literalist fundamentalists), but the *mechanism* is. The intelligent designers disagree with the Darwinian *mechanism*. What is important about this connection between intelligent design and creationism is that the tension or "battle" is not between *evolution* and creationism, but between *Darwinism* and creationism; the acceptance of evolution does not guard against incorrect, creationist, views.

b. Evolution through punctuated equilibrium

In the late twentieth century the only biologist whose fame could rival Dawkins' was Stephen Jay Gould. Gould is best known for his theory of evolution through "punctuated equilibrium", by which evolutionary change occurs rapidly, punctuating long periods of relative stasis (Gould, 2002, p. 766). Gould disagrees with the neo-Darwinian claim that "living organisms exist for the benefit of DNA rather than the other way around" (Dawkins, 1986, p. 126) and argues instead that "species [are] the basic units or atoms of macroevolution" (Gould, 2002, p. 781).

This distinction between "macro" and "micro" evolution is the central disagreement that Gould has with Darwinism. Whereas Darwinism argues that macro-evolution is simply the observation of micro-evolutionary change over a long period of time (marked by the disappearance or extinction of "intermediaries" [Darwin, 2009, p. 165]), punctuated equilibrium argues that micro-evolution (i.e. minute variations) is largely arbitrary and does not change the species, until such time as macro-evolution (i.e. observable speciation) occurs (Gould, 2002, p. 787).

The importance of this cannot be overstated. The assertion that "*natura non facit saltum*" (and the role of accumulation for which it argues) is absolutely central to the Darwinian theory and Gould is clear that "punctuated equilibrium makes the strong claim that in most cases, effectively no change accumulates at all" (Gould, 2002, p. 767). What is somewhat ironic about this disagreement is that, while punctuated equilibrium is characterized by relatively large periods of stasis, it actually strongly disagrees with the primacy of preservation. It fails to acknowledge Monod's simple but crucial observation that without the prior presence of preservation, change cannot be "caught" by the organism and then replicated. Furthermore, without this primacy of preservation, punctuated equilibrium has to disagree with Dobzhansky's observation that all variation is identical to mutations caused in a laboratory through provoking genes to malfunction with x-rays. Gould's theory is based on the relative sparsity of the fossil record (Gould, 2002, p. 766), and this can be explained by other reasons (such as the fact that not all organisms are fossilized or that some fossils are eroded [Williams, 2001, pp. 71–2]).

This does not mean that Gould's contribution to evolutionary theory must be completely dismissed. If all Gould means here is that evolutionary change is not smooth and regular and that the difference between Darwinism and punctuated equilibrium is "an arbitrary choice of the timescale" (Miller, 1999, p. 118), then he may be making a valid point. If, however, punctuated equilibrium stands by Gould's claim that it "is a theory about the evolution of phenotypes" (Gould, 2002, p. 810) and that "no change accumulates at all" (Gould, 2002, p. 767), then it must be rejected.

c. Convergent evolution

Another important evolutionary biologist presenting a novel theory of evolution is Simon Conway Morris, who argues for "convergent evolution". The essential point that Morris makes is that there are only a limited number of "solutions" to the problem of survival, and these solutions have been "found" many different times, and quite independently, by many different organisms. Morris writes that "despite different starting points, different lineages often converge on particular destinations" (Morris, 2003, p. 124). This has led to his famous claim that were the evolutionary tape replayed, evolution would transpire in more or less the same way (Morris, 2003, p. 106).

It is obvious that this provokes the conclusion that evolution is directed in some way. By arguing in his *Life's Solution: Inevitable Humans in a Lonely Universe* that there are only certain "solutions" to problems, Morris claims that "this book aims, if nothing else, to refute the notion of the 'dominance of contingency'" (Morris, 2003, p. 297). This is a notion that this chapter has shown is crucial to neo-Darwinism, and for which many geneticists have provided evidence.

At face value, it appears that Morris is confusing genetic mutation and natural selection. It may very well be, for example in certain environments, that only certain "solutions" are possible (e.g. a thick coat in the Arctic), but before these solutions are selected they must appear; Darwinism is clear that the environment cannot provoke genetic mutation (which is entirely accidental), let alone provoke a particular mutation. If Morris means that some things are *unviable* rather than *unavailable* (i.e. his point is about selection rather than mutation), then it can be salvaged. However, even this does not guarantee that the tape will replay in more or less the same way.[6]

Another criticism of convergent evolution is that by describing evolution as a "search engine" in search of "solutions" (Morris, 2010, p. 151), Morris disagrees with the central Darwinian claim that evolution is about survival and not perfection. As Julian Huxley notes, natural selection "does not ensure progress, or maximal advantage" (Huxley, 1942, p. 466). Certainly, some mutations will always provide the organism with an advantage in certain environments (such as swim bladders in the ocean), but this in no way implies that evolution actively seeks these solutions. It would be a disservice to Morris to pigeonhole him with the "intelligent designers"; however, the same arguments that were levelled against them can also be used to criticize Morris.

d. Epigenetics
It is also important to mention another area of evolutionary research that is becoming increasingly influential: epigenetics. Epigenetics is significant as some have claimed that it provides evidence for Lamarckism and could lead to a revival of that theory. Essentially, epigenetics refers to the chemical and hormonal environment in which DNA is replicated and the possible effect it has on that replication. Epigenetics argues that

certain characteristics can be replicated faithfully due to a consistent chemical environment. Nessa Carey explains that "wherever possible, a cell's default setting is to maintain the genome in its original state, as much as it can" (Carey, 2012, p. 264), yet whereas this is achieved because "the epigenome is usually reset in sexual reproduction . . . this process is occasionally subverted to allow the inheritance of acquired characteristics" (Carey, 2012, p. 307).

However, Carey continues that the "occasional subversion" of this epigenetic reset is a "predominantly random process" (Carey, 2012, p. 93). In this way, while epigenetics might be an important area of evolutionary research, the fact that it is entirely random means it must be Darwinian; the presence of acquired characteristics is entirely random and accidental. Warren Burggren comments that these epigenetic mechanisms are "themselves heritable and selected for" (Burggren, 2014, p. 685). Acquired characteristics (the central feature of Lamarckism) may be a real phenomenon due to epigenetics, but this does not represent a return to Lamarckism; it can still be explained by Darwinian sensibilities.

What is the neo-Darwinian synthesis?

The neo-Darwinian synthesis is essentially comprised of two distinct components: variation caused by accidental genetic mutation and blind natural selection (or, the observation of differential survival). Evolutionary change happens as a result of these two elements.

By genetic mutation is meant the accidental and unintentional change in genetic information that occurs when there is a mistake or "deleterious effect" in replication. The use of rhetoric such as "deleterious" does not imply that such an effect is negative. Neo-Darwinism is emphatic that judgement of genetic mutation is entirely neutral or subjective. Likewise, the use of the word "accidental" does not involve negative connotations. Instead, it is used to denote that the change in genetic information that occurs due to imperfect replication is unintentional. All mutations are mistakes, but not all mistakes have negative consequences.

When mutation is described as random, this does not indicate that an infinite number of possible mutations exist; rather, that it is impossible to predict when a mutation will occur and that it is impossible to predict which particular mutation will occur.[7] Genes will continue to replicate faithfully until a mutation occurs, at which point that particular mutation will replicate until another mutation occurs, and so on. Hence the importance and primacy of accumulation.

By natural selection is meant the observation of differential survival of individuals due to competition over limited resources, and their resulting ability to reproduce. This differential survival gives the impression of a selection by nature that is comparable to the selective breeding that is carried out by humanity in order to emphasize a particular trait. The difference is that, whereas artificial selection is a deliberate interference in reproduction with the prior intention to emphasize a particular trait, natural selection is simply the preservation of those individuals that, through random accident, have been afforded variations that allow them to survive and reproduce at a better rate than others.

Whereas artificial selection is an active interference, natural selection is simply a passive observation. It is in this sense that natural selection can be said to be blind. However, whilst it may be blind, it is not random; the particular occurrence of a specific mutation is entirely random and accidental, but if this affords the individual a survival or reproductive advantage then it will, probably, survive long enough to reproduce—its survival and reproduction are not random but its appearance in the first place is by the random mutation. Thus, Dawkins can write that evolution is the accumulation of "random hereditary changes" by "non-random survival" (Dawkins, 2004, p. 92).

Whilst the neo-Darwinian synthesis is the interaction between these two principles, they do not have a causal relationship: genetic mutation does not influence selection "criteria", and natural selection cannot cause a mutation to occur, let alone a particular mutation. Instead, as Dawkins writes,

> [n]o matter how strong a potential pressure may be, no evolution will result unless there is genetic variation for it to work on. (Dawkins, 1999, p. 42)

Natural selection does not have a causal effect on mutation, which is random and unintentional. If no mutation occurs, then no particular selection can take place. If the individual cannot survive in its environment, then it will become extinct. This, however, does not imply that there are particular traits that have objective value; the correct mutation is simply the one that affords survival.

However, the most important element of the neo-Darwinian synthesis is that the whole "process" is concerned more with preservation than with creation. The evidence, with reference to Dobzhansky, strongly suggests that genetic mutation occurs when a particular gene fails to replicate properly. This provokes the conclusion that mutation is a mistake and represents, not a moment of creation, but a moment of failure of preservation. True, this mutation may provide the individual with a particular survival advantage, but any advantage given is purely accidental. Such an interpretation is strengthened by the fact that genes have mechanisms to prevent this change (see Guttman, Griffiths, Suzuki and Cullis, 2002, pp. 123–4 and Birdsell and Wills, 2003, pp. 120–1) and that natural selection is concerned with further preventing this change (Dawkins, 1986, p. 125). Likewise, a proper understanding of Darwin's theory of natural selection suggests quite strongly that what is happening is not an active selection of perfect forms—it is not a "search engine"—but the preservation of those that are able to survive and reproduce.

Evolutionary change occurs when there is a mistake in the continuing and relentless copying of genetic material, and this mistake is itself able to replicate. This principle of continual copying and replication of an imperfect copying process is the fundamental precept that this chapter concludes: change is explained *only* as the result of the imperfect nature of that replication.

Theological and philosophical considerations

By way of conclusion, the ideas that this chapter has argued in favour of will be brought together in the context of their implications for theology. As this book is concerned with theology rather than biology, it is important to outline a number of principles that will be used to re-think and re-interpret theology and, more specifically, Christology.

a. God does not control evolution

In *Chance and Necessity*, Jacques Monod notes that there are two distinct ways in which theologians and philosophers claim that God directs evolution, which he calls "vitalism" and "animism". "Vitalism" is the idea that God has a continuing control over the direction of evolution, whereas "animism" is the idea that God set up evolution with certain values that act as a goal or *telos*, pulling and attracting evolution towards them, e.g. consciousness (Monod, 1972, pp. 32ff.) or complexity. The first could be seen as being supported by traditional theists, whereas the second is supported by traditional deists. Yet neither of these ideas are supported by neo-Darwinism.

It is often argued, and rightly so, that an absence of evidence is not evidence of absence. Just because science (and more specifically biology) cannot detect a divine influence—it cannot find any evidence of a guiding hand on evolution—it does not mean that there isn't one; it simply means that science is unable to detect it. However, importantly, this is not the picture that neo-Darwinism paints. Neo-Darwinism does not remain neutral on the question of divine influence; it is not that there is an *absence* of evidence but that there *is evidence* of absence.

The lynchpin of this chapter has been to show that, contrary to what theologians claim, evolution is more concerned with preservation than with change (or creation). In fact, change *only* happens when genetic replication malfunctions and such malfunction is actively resisted by genes. Dobzhansky's observation—that all laboratory-provoked variations caused by purposefully disturbing the gene's ability to replicate are exactly the same variations that appear in nature and are the raw material of natural selection—is sufficient to repudiate the idea that God controls genetic mutation (see also Fisher, 1930, p. 20).

Of course, drawing on what Monod called "animism", it could be argued that this "tendency" to malfunction is something designed by God in order to create for God. However, this is denied for two reasons: (a) because there are *no* values (not even consciousness) that enjoy an objective advantage and (b) because genes actively resist and attempt to thwart this malfunction and change (as Dawkins [2006, p. 86], Monod [1972, pp. 108–9], and Guttman *et al.* [2002, pp. 123–4] have all acknowledged). Mutation is completely accidental, actively resisted, not always creative, and not always positive. Mutation is not provoked by any survival need and only accidentally provides the organism with a slight survival or reproductive advantage.

In any case, however, regardless of whether God causes variations or not, if they are not selected, then they will not survive and thrive. Any theologian who claims that God can cause and influence genetic mutation still "needs to account for how God might ensure the subsequent natural selection of divinely mutated progeny" (Saunders, 2002, p. 114).

When it comes to natural selection, there is also evidence against the idea that it is controlled by God. As Darwin argues, and Williams (2001, p. 137) and Crysdale and Ormerod (2013, p. 36) emphasize, natural selection is *not* creative. Blind natural selection can only eliminate, or at best modify—it cannot provoke the creation of something novel. This is supported by the idea of form before function, which argues that anything that has evolved has done so through putting to different function something that is already present.

The concept of divine direction of natural selection is also rejected on the basis of an absence of objective value. Natural selection does not have a single, objective set of criteria through which it filters individuals. As Julian Huxley so matter-of-factly notes, "organisms are selected, not on the basis of conformity to an ideal plan, not in relation to complete functional efficiency, but on the basis of survival" (Huxley, 1942, p. 449), and so "all that natural selection can ensure is survival. It does not ensure progress, or maximal advantage, or any other ideal state of affairs" (Huxley, 1942, p. 466). What survives does so for no other reason than it has survived. Such an observation led Dawkins to call natural selection a tautology (Dawkins, 1999, p. 181; see also Berry, 1982, p. 57). However, perhaps the biggest evidence against the notion that God can control or direct selection is that natural selection is not a real phenomenon to begin with. Natural selection is only an observation that some organisms survive longer and reproduce more ancestors than others due to competition over limited resources. Once there is variation in the population (due to mutation), then this variation has an effect on survival or reproduction, and will influence which genes are able to replicate better.

b. Finished versus unfinished creation

One of the important implications of this denial of divine influence, which is strengthened by the denial of objective values against which is judged the relative worth of individuals (or what can be called the teleological neutrality of evolution), is the rejection of a tension between finished and unfinished creation. Evolution is not controlled by God and does not, as Morris argued, search for certain solutions or perfections. The universe is not currently in an unfinished state, full of "nascent transitional forms" (Smith, 1988, p. 7) that will eventually be finished through the process of evolution. The teleological neutrality of evolution argues there is nothing towards which mutation and selection is proceeding, and no point, therefore, at which evolution can be said to be complete. The universe neither comes into being finished, nor will it be finished in the future.

This rejection of such a tension provokes an important conclusion: the neo-Darwinian paradigm is permanent. Neo-Darwinism does not describe a temporary process by which things come into being; rather it describes a permanent condition of what it means to be a creature. Neo-Darwinism is not a temporary condition of the universe, but a permanent one.

b.1. Universal Darwinism

The permanency of neo-Darwinism also has another facet: it is not just a biological phenomenon. Although there are some who criticize the application of Darwinism outside of biology (see McGrath, 2005, pp. 121ff. and Foster, 2009, pp. 102ff.), many commentators have noted that Darwinism has a much wider scope than biology. Mivart suggested that Darwinism could be applied to the "stable equilibrium of the Solar System" (Mivart, 1871, p. 22). Likewise, Corte notes that Darwinism can be applied to chemistry, noting that "heavy elements" evolved from "lighter ones" (Corte, 1960, p. 60). Hans Jonas, agreeing with the interpretation of this chapter that "the foundation of all order in nature, of any nature at all, lies in conservation", tells the story of the universe from nothing to complex life from a neo-Darwinian perspective, indicating that neo-Darwinism has a far wider scope than biology (Jonas, 1996, pp. 168–9).

Dawkins has also more famously argued that neo-Darwinism has a wider scope than biology: he claims that neo-Darwinism can be applied to the social sphere, with what he calls "memes". In arguing for this idea Dawkins says:

> The real unit of natural selection was any kind of *replicator*, any unit of which copies are made, with occasional errors, and with some influence or power over their own probability of replication. The genetic natural selection identified by Neo-Darwinism as the driving force of evolution on this planet was only a special case of a more general process that I came to dub 'universal Darwinism'. (Dawkins, 2004, p. 149)

What is described in this chapter, therefore, is not restricted or limited to biological evolution. Darwinism is far more inclusive and far more relevant. What is offered here, then, is not how a biological principle can have implications for theology, but how that biological principle expounds a far wider and more general principle. Neo-Darwinism does not simply describe how things *came to be*; it describes how things *are*.

c. Ontology

All of this points to perhaps the most important theological implication of neo-Darwinism: that evolution cannot be understood as simply a replacement for the Genesis narrative. Neo-Darwinism doesn't *only*, nor even *primarily*, describe how things come into being (indeed, neo-Darwinism doesn't think that anything *new* comes into being [i.e. is created]—everything is "just" a mutated version of something else). Rather, it primarily describes how things *are*. In other words, the point of connection between evolution and theology—the point where theology and evolution overlap—is not creation, as many (if not all) theologians suppose, but ontology.

This, of course, does not deny the purpose for which the neo-Darwinian synthesis was originally devised (i.e. to explain the diversity in the world), but to see this as the primary element of neo-Darwinism is to miss what this chapter has argued—the creation of new species is an accidental by-product of the malfunction of genetic replication.

It is generally argued that neo-Darwinism favours what has been termed "becoming" over and against "being". The idea of a static "being" is generally applied to those philosophers who disagreed with the mutability of species and, instead, argued for a fixity of species. Plato is a perfect example of this view. By postulating a realm of ideals or forms that do not change, Plato claimed that *what is real does not change*; real being is unchanging.

When the mutability of species is accepted, it is obvious that such a static ontology must be re-evaluated. "Being" can no longer be understood as unchanging. The obvious mutability of the universe is evidence that what is real is changing. This led philosophers and theologians to reject "being" in favour of "becoming" and arguing that *what is real, changes*. Thus, Haught writes:

> A metaphysics of *esse* (or "being") [has] obscured the obvious fact of nature's constant "becoming" and its perpetual movement toward the future. (Haught, 2000, p. 84)

Nothing remains the same, everything changes, and what is real must be that which changes. Moreover, as Haught clearly evidences, those who support this idea of "becoming" see it as progressive and future-oriented.

However, as this chapter has argued, this idea is not correct either. While neo-Darwinism is certainly a theory postulated to explain change, it explains this change through the primacy of preservation. Evolution is about preservation, not change; change only happens when preservation is not perfect. To put this differently, evolution is not *dynamic*, it is *mutable*; evolution does not promote change, but change happens to it. Evolution is not forward-looking—it is backward-looking. Evolution is not about what organisms change *into*; it is about what they change *from*.

This means that, from a neo-Darwinian perspective, neither "being" nor "becoming" can adequately explain what is real. Certainly, it is undeniable that there is "becoming", but not in the dynamic, future-oriented sense that other philosophers and theologians understand it. Indeed, this chapter has argued that change, or "becoming", is actively resisted; if what is real is that which changes, then this should not be resisted. What is real, then, *is what attempts to stay the same whilst being unable to do so.*

This could be termed "half-fixity", which emphasizes the primacy of the fixity of species yet acknowledges that they never remain fixed. Or, to put it differently, this term accepts that the fixity of species is incorrect, yet also guards against the opposite claim that there is a "perpetual movement towards the future". However, with an eye towards the role that imitation will play in subsequent chapters, this ontology is better described as an ontology of "malfunctioning replication" or "imperfect copying". What is real is that which preserves, but that preservation is never absolutely perfect.

CHAPTER 2

Theological Anthropology

The neo-Darwinian synthesis, by adequately explaining all diversity in the universe, has profound implications for doctrines of humanity. While Darwin himself was careful to leave out references to the genesis of humanity in his *Origin*, there was no mistaking that inferences could clearly be made.

Superficially, Darwin's theory (along with Charles Lyell's geological estimations of the age of the earth, which heavily influenced Darwin) clearly disagreed with the anthropological history outlined in the book of Genesis, which had been almost universally accepted by the Abrahamic West. However, neo-Darwinism does not merely disagree with the historical accuracy of the Genesis narrative. By making comments about human origins, Darwinism has important things to say about both human relationships with other, non-human creatures[8] and with God.

Firstly, the neo-Darwinian synthesis disagrees with the biblical role of humanity in terms of its inherent superiority over other creatures. Humanity, by sharing an origin with other creatures—indeed, an origin *in* other creatures—cannot be thought of as transcending other creatures. The relationship of humanity with the rest of creation must now be thought of in terms of equality, not superiority. Secondly, the neo-Darwinian synthesis, by disagreeing with the historical accuracy of the Genesis narrative, leaves no room for an original perfection and subsequent Fall of humanity.

In this way, Darwinism is much more optimistic about humanity's relationship with God. Importantly, this is not a denial of original sin, as many commentators on theological interpretations of evolution would claim is necessary. Neo-Darwinism does not necessarily disagree with what theologians have said about human openness to failure and mistake. However, it disagrees that this is a *temporary* feature of human experience; there is no original or final perfection that is not characterized by openness to failure.

This, it will be argued, is demanded by, and thus represents a connection with, the neo-Darwinian synthesis. Furthermore, and drawing on the conclusions of Chapter 1, this openness to failure does not just characterize human experience but the experience of all creatures. In other words, sin is not flawed and damaged relationship with God but the only possible way that relationship with God can exist; if creatures were not mutable (i.e. open to failure) they would be God and not creatures.

The central tenet of this chapter, therefore, is the rejection of human uniqueness. Humanity is no longer unique when it comes to the created universe in which it finds itself, differing only quantitatively, or in *degree*, from all other creatures not qualitatively, or in *kind*. There is a much greater unity and inclusivity about the universe than biblical anthropology supposed. Likewise, humanity is no longer unique in terms of its relationship with God. By denying its superiority over other creatures, coupled with the rejection of a negative interpretation of sin and denial that sin uniquely describes human relationship with God, neo-Darwinism argues that God does not treat human creatures any differently than other creatures.

This chapter will disagree with the uniqueness of humanity in two important ways. Firstly, it will argue that the traditional, biblical understanding of the primacy of humanity over the rest of creation must be rejected. Neo-Darwinism disagrees with a literal interpretation of the Genesis narrative and, in doing so, disagrees with the role and place that humanity occupies in the universe, especially in terms of its relationship with non-human creatures.

Secondly, this chapter will disagree with human uniqueness in terms of original sin. While traditionally the doctrine of original sin is understood as a punishment conferred on humanity as the result of a historical transgression, the neo-Darwinian synthesis strongly disagrees with this. Rather, by seeing original sin as something that is a natural and inherent part of being created, and finding parallels with the neo-Darwinian ontology that was outlined in the previous chapter, original sin must be something that is descriptive of all creatures, rather than uniquely human.

The rejection of human primacy

For the most part, at least up until the publication of Darwin's groundbreaking and paradigm-shifting theory of natural selection, theological anthropology has been an almost exclusively biblical endeavour. The opening chapters of Genesis, along with a number of other passages, formed the basis of how Christian theology understood what humanity is and what its relationship to God and the rest of the universe is. This understanding can be summarized as arguing for human primacy. This primacy claims that humanity is the *reason* for creation, that the rest of the universe is created for humanity, and, therefore, humanity enjoys a uniquely close relationship with God (Hall, Rae and Holmes, 2010, p. 280).

The opening chapters of Genesis present two distinct accounts of the creation of the universe. In the first account, God starts with the "heavens and the earth" and proceeds through light, sky, sea, and non-human animals before culminating in humanity. The implication of creating humanity last is that it is the most important. Humanity is the crowning achievement of God's creation and, thus, is afforded a special relationship with God. This special relationship with God also means that humanity has a particular relationship with the rest of the universe. This relationship is one of dominion, which, whilst not explicitly defined in Genesis, is normally taken to mean one of two things. Some commentators have argued that the universe has been created exclusively for humanity, which can, therefore, take advantage of it for its own sake (see Hoekema, 1994, p. 88). Other commentators take a different view of this dominion, arguing instead that it implies a responsibility for taking care of the universe that other creatures do not have.

In the second account of creation in Genesis humanity is created first (after the heavens and the earth), and humanity is tasked with naming the rest of creation as it is created by God. This means that humanity enjoys a primacy over creation. What is more explicit about this second account (only somewhat implied in the first account) is that the rest of the universe is created *for* humanity, and its worth is only judged in relation to its value to humanity.

Thus, C. S. Lewis writes that "the beasts are to be understood only in their relation to man, and through man, to God" (C. S. Lewis, quoted in Southgate, 2008, p. 11). The sole role of the rest of the universe (including, for the Genesis author(s), women), and thus the sole reason for their being created, is to placate the loneliness of man (Genesis 2:18). The identity of humanity is construed in its relationship with God, whereas the identity of all other creatures is construed in their relationship with humanity. In this way "everything else finds its completion in the things that have been made, but human beings find their fulfilment in God" (Wilken, 2003, p. 150).

In both accounts of creation, therefore, humanity is seen as being God's helper in some way (Phillip Hefner calls humanity "created co-creators" [Hefner, 1993, pp. 23ff.]). In both accounts, humanity is afforded a responsibility in and to creation that places it higher than the rest of creation, a responsibility that makes it unique. In the first account, humanity is seen as the crowning achievement of creation, whereas in the second account the rest of creation is explicitly created solely for humanity, and thus can only be understood within the context of the primacy of humanity.

There are other instances where these creation myths are expounded. Psalm 8 also reiterates this notion of the special place that humanity has in creation. Of course, the Psalms are more prayerful in nature and thus lend themselves more to an allegorical rather than a literal interpretation, but, nonetheless, they give an insight into the theology of the time. In this instance, the psalmist writes:

> When I look at your heavens, the work of your fingers, the moon
> and the stars that you have established; what are human beings

that you are mindful of them, mortals that you care for them? Yet you have made them a little lower than God, and crowned them with glory and honour. You have given them dominion over the works of your hands; you have put all things under their feet ... (Psalm 8:3–6)

The language used by the psalmist is almost identical to that of the author(s) of Genesis (thus suggesting a devotional intention behind the Genesis accounts, rather than strictly historical). Humanity is afforded pride of place in creation, and that pride is understood as dominion. The psalmist, however, goes slightly further than the Genesis author(s) by explicitly claiming that humanity is only slightly lower than God. This is almost certainly the intention behind the Genesis accounts, particularly in relation to being created in the "image of God", yet the psalmist makes this implication explicit.

a. Evolutionary anthropology

It is interesting that incorrect theories of evolution tend to support biblical anthropology, and thus the place of humanity in the universe that biblical anthropology supports. Perhaps the most famous example of this agreement is the evolutionary theology of Pierre Teilhard de Chardin. Briefly, Teilhard argued that evolution not only represented an increasing complexification of creation, but that such a complexification was accompanied by a correlative material convergence and increase in spirituality (Teilhard de Chardin, 1959, p. 60).

Teilhard also argued that, at certain moments in history, such complexification and spiritualization reach a point where something entirely new emerges and breaks forth. These include the moments when life emerged from inanimate matter and when consciousness became a dominant force in creation. According to Teilhard, there is a qualitative difference between life and non-life and, comparatively, there is a qualitative difference between life and consciousness. Such a difference is understood as an irreversible progress and improvement of creation. It is in this direction that Teilhard writes that "life is the rise of consciousness" (Teilhard de Chardin, 1959, p. 153), and that "evolution is an ascent towards consciousness" (Teilhard de Chardin, 1959, p. 258).

Such an understanding of evolution led Teilhard to claim that animals "are separated from [humanity] by a chasm" (Teilhard de Chardin, 1959, p. 166). The difference between humanity and the rest of creation is not one of degree but one of nature; there is an ontological difference between the universe and humanity, which enjoys a closer relationship with God as a result. It is clear that such an understanding of humanity follows the biblical anthropology outlined above, particularly the first account of creation. As creation unfurls it becomes more and more complex, eventually culminating in humanity.

—

However, such a directed interpretation of evolution has already been completely rejected. Neo-Darwinism "knows of no teleology" (Deane-Drummond, 2009, p. 229). While it agrees with the first Genesis narrative regarding the order in which the things in the universe came into being, it disagrees that such an order implies an inherent judgement value. Every creature that has survived for a significant period of time has "equally solve[d] life's problem", which means that it is not possible to claim that "one form of life is superior to another" (Kenney, 1970, p. 65). Life is not a "Christmas tree, with homo sapiens accorded a pseudo-angelic position" (Peacocke, 2001a, p. 27); rather, life is a "bush" or a "shrub" (Dodson, 1984, p. 137) that has no particular direction.

This subjective approach to judgement has already been acknowledged as an important and crucial element of Darwin's theory. There is no objective way of judging which form of life is superior, as the criteria that judge success are themselves completely subjective: "Fitness . . . is qualitatively different for every different organism" (Fisher, 1930, p. 37). Such criteria are always short term (Dawkins, 1986, p. 50) and constantly changing (Cole-Turner, 1993, p. 43). Indeed, it is entirely possible that "all complex forms of life will go extinct and only the microbial prokaryotic cells will survive" (Peters and Hewlett, 2003, p. 119). Thus, Darwin writes that biologists have not agreed "what is meant by high and low forms" (Darwin, 2009, p. 297).

Not only does the subjective approach to judgement question the primacy of humanity, but the prevalence of accumulation also has an important role. Precisely because all life on earth is connected by continual and miniscule genetic variation—and nature makes no leaps—it is impossible to demarcate with anything approaching precision where humanity begins and other creatures end. Darwin himself reached precisely this conclusion. Darwin argued that the term "species" was "arbitrarily given for the sake of convenience" (Darwin, 2009, p. 56) and that "there is no infallible criterion by which to distinguish species and well-marked varieties" (Darwin, 2009, p. 59). This means, Darwin reasoned, that "numberless intermediate varieties, linking most closely all the species of the same group together, must assuredly have existed", but natural selection has "exterminated" those intermediate links (Darwin, 2009, p. 165). Thus, Darwin writes that:

> If every form which has ever lived on this earth were suddenly to reappear, though it would be quite impossible to give definitions by which each group could be distinguished from other groups, as all would blend together by steps as fine as those between the finest existing varieties, nevertheless a natural classification, or at least a natural arrangement, would be possible. (Darwin, 2009, p. 378)

This is so important for neo-Darwinism. The separation of individuals into variations or species is only possible because the "intermediaries" have become extinct, and this must conclude that there is absolutely no objective and ontological distinction and separation of humanity from other non-human creatures. Between every individual creature that exists now—and has ever lived or existed[9]—there is a familial, genealogical link.[10]

Richard Dawkins makes the same point:

> To a non-punctuationist, "the species" is definable only because the awkward intermediates are dead. An extreme anti-punctuationist, taking a long view of the entity of evolutionary history, cannot see "the species" as a discrete entity at all. He can see only a smeary continuum ... the extreme anti-punctuationist sees "the species" as an arbitrary stretch of a continuous flowing

river, with no particular reason to draw lines delimiting its beginning and end. (Dawkins, 1986, p. 264)

Dawkins' criticism of Stephen Jay Gould here is clear; however, the important point is the "smeary continuum". All individuals are related to all other individuals, and there is nothing that can be pointed to that can separate humanity from the rest of creation. Humanity is not only truly a part of the universe from which it came, but it cannot be differentiated from it precisely because of that fact. Humanity cannot be separated from the rest of creation, which means that there is nothing so concrete as human nature. In fact, the whole idea of distinct natures in general is completely alien to neo-Darwinism. Moritz agrees that what constitutes human uniqueness and, indeed, human nature itself, has become problematic after the development of the neo-Darwinian synthesis (Moritz, 2015, p. 45).

There is another important point to make here regarding neo-Darwinian anthropology, namely, the accidental nature of humanity. Just as there is no direction or teleology in evolution (which suggests that humanity must be afforded a higher value) so there is nothing about evolution that suggests humanity is guaranteed to evolve. This accidental nature of humanity's evolution has already been touched upon in relation to Simon Conway Morris' assertion that the "tape of evolution" would always produce intelligent, hominid life, and the rejection of this occurring also serves to reject the uniqueness and value of humanity. The possibility of humanity evolving is extremely improbable, and subject to historical accident.

For many, this could be seen as a relegation of humanity, a "lowering" of it to the level of the beasts, but this is not necessarily the case. There is nothing to deny the opposite claim that all other creatures have been promoted to enjoy the status that humanity enjoys. If there is nothing that can adequately distinguish humanity from the rest of creation, then any theological conclusions must be reformulated; nothing theological can now be said about humanity without necessarily including all other creatures.

This promotion of all creatures to the level of humanity, together with (and made necessary by) the genealogical link between all creatures, means that there is now an inherent unity between all creatures. Drawing on the inclusive attitude of Francis of Assisi, Coulson remarks that the locust is our brother and the bacteriophage our sister (Coulson, 1958, p. 125).[11] Leonardo Boff (who was formally a member of the Franciscan Order) also sees all creatures as sharing a familial unity: "all living things are brothers and sisters" (Boff, 1997, p. 211).[12]

Neo-Darwinism certainly has its critics, but there can be no doubt that it is much more inclusive than biblical anthropology. The Bible claims that "there is no longer Jew or Greek, there is no longer slave or free, there is no longer male and female" (Galatians 3:28), yet neo-Darwinism goes further. For the neo-Darwinist there is no longer human and non-human and there is no longer living and non-living—all are united.

b. The image of God
No discussion of theological anthropology could be complete without a treatment of the image of God. It would not be too radical to claim that being created in the image of God is the central element of theological anthropology—the claim that holds all the others together. For the biblical authors, the uniqueness of humanity, that which separated it from the rest of creation, was that humanity alone was created in the image of God. Although the Bible suggests that humanity images God by having dominion over the rest of creation (i.e. that in the same way that God has power and responsibility over humanity so humanity has power and responsibility over other creatures), there has been much speculation over the centuries as to what constitutes this image.

At some point in the history of theological speculation, and in various ways, the possession of self-consciousness, rationality, episodic memory, language, morality, freedom, the ability to form relationships, and the capacity for religion have all been postulated as representing what it means for humanity to be created in the image of God. It is not necessary to discuss every aspect of these theories: if it can be shown that other non-human creatures also possess the capacities that some have claimed represent human uniqueness, then the conclusion must be reached that being created in the image of God does not argue for human uniqueness, and if being created in the image of God provides humanity with a closer relationship with God, then this close relationship must be shared with other creatures.

—

These days, nobody seriously doubts that non-human creatures are conscious. Descartes famously argued that non-human animals are simply automatons that only appear to be conscious. At the opposite position, Teilhard de Chardin, among others, represents the opinion of panpsychism, the view that all matter has a degree of consciousness corresponding to its material complexity. The problem with both of these ideas is that they separate consciousness/mind from the brain. For Descartes, consciousness is denied when a brain is present; for Teilhard, consciousness is affirmed when a brain is absent.

However, there is simply too much evidence to ignore the conclusion that brain states are entirely responsible for, and solely sufficient to explain, consciousness. Thus, "there are no 'mental' events that are without a physical realization in the brain" (Murphy, 1998, p. 2). It is perhaps too simplistic to claim that there is a close "one-to-one" correlation between consciousness and brain states, but this does not mean that the brain cannot be entirely responsible for the mind and consciousness.

However, while there is no doubt that non-human animals are conscious, self-consciousness—as Teilhard de Chardin puts it, "knowing that you know" (Teilhard de Chardin, 2004, p. 126)—is often only ascribed to humanity. Only humanity has the ability to view the world from a "first-person" perspective and understand that it is different. Yet, this is not necessarily the case, as biological and neurological research has shown. Experiments such as the "mirror test" (which tests whether animals can recognize themselves in a mirror) have suggested that higher apes, elephants, dolphins and magpies all display behaviour that could demonstrate the possession of self-consciousness (see, for example, Prior, Schwarz and Güntürkün, 2008). Pigs and crows also seem to be able to use a mirror to find food; if this does not demonstrate self-consciousness, it certainly demonstrates the ability for abstract thought and rational thinking. If self-consciousness and rationality are being made in the image of God, then humanity can no longer be considered unique.

Closely tied to the possession of consciousness and rational thought is the possession of an episodic memory, which has been proposed both as the image of God (Augustine, 1991a, pp. 301–2) and absent in non-human animals (Brown, 1998, p. 116). However, once again, behavioural experiments seem to suggest otherwise. Crows, for example, have been shown to delay gratification by refusing a treat if they think they will be offered a better one later (see, for example, Hillemann, Bugnyar, Kotrschal and Wascher, 2014).

The possession of language is also a point of contention, being claimed by some as "the greatest chasm between the most intelligent non-human primates and human beings" (Brown, 1998, p. 104; see also Williams, 2001, p. 120). Yet, again, plenty of non-human creatures have displayed language. It has even been suggested that dolphins have such a complex language it could demonstrate a rudimentary grammar and sentence formation (see, for example, Herman, Kuczaj and Holder, 1993).

All other capacities that have been proposed as representing the image of God can be explained by the possession of rational thought, language, and self-consciousness. When an individual creature is able to use memories of past behaviour to influence the way it thinks about its current situation and how it solves problems, then things like freedom, moral behaviour, and the ability to form and nurture relationships are inevitable.

Admittedly, the ascription of rational thought and self-consciousness to non-human animals still excludes the majority of creatures, which, while it does deny human uniqueness, still does not demand that *all* creatures are made in the image of God. However, the point here is not to ascertain which creatures are made in the image of God and which are not. The point is to show that human uniqueness must be rejected: humanity is no longer separated from non-human creatures by a chasm, and there is an ontological unity among all creatures.

Those capacities that have at one time or another been suggested to demonstrate human uniqueness only differ in *degree* from other creatures, not in *kind*; there is a *quantitative* difference between humans and non-humans, not a *qualitative* difference. The point is not to show that all creatures are self-conscious, but to show that self-consciousness does not separate humanity and non-human creatures. This is a necessary conclusion based on the primacy of accumulation for neo-Darwinism. This must mean that *all* creatures are made in the image of God, not because all creatures display identical behaviour in some respect, but because there is a genealogical, familial, ontological link between all creatures.

c. The soul

There is another claimant for the image of God that needs to be considered: that humanity uniquely possesses an immaterial soul that separates it from the rest of creation and places it in a closer relationship with God. Humanity, it is claimed, is composed of both a material body and an immaterial soul, and it is the possession of a soul that separates it from other non-human creatures.

However, this dualist understanding must be rejected. To begin with, there is no clear biblical support for the soul (Anderson, 1998, p. 179). Although the presence or absence of a particular idea in the Bible is not crucially important (this chapter has already criticized biblical theology), it still contributes to the rejection of the soul. If there is no biblical reason to retain the idea, then it makes other criticism of it easier to take.

Thus, the previous chapter on neo-Darwinism was clear that the general theory of cosmological dualism must be rejected. The evidence that there is no direction to evolution and that all mutation can be adequately explained through material causes means that there is no evidence for an immaterial influence on the universe. Cosmological dualism is just not necessary. This rejection of cosmological dualism means that the more specific theory of anthropological dualism must also be rejected. There is simply no evidence that the soul exists.

Perhaps more importantly, any capacity that is attributed to humanity on the basis that humans possess a soul is found to have other causes. Murphy is clear that "nearly all of the human capacities or faculties once attributed to the *soul* are now seen as functions of the brain" (Murphy, 1998, p. 1). The most common of these faculties is intelligence (see Hall, Rae and Holmes, 2010, p. 286 [quoting Gregory of Nazianzus] and Aquinas, 1998, p. 428). The separation of consciousness and mind from the brain has already been noted to be incorrect. Cooper confirms that "without certain brain functions, there is no conscious mind or personality whatsoever" (Cooper, 1989, p. 22). There is simply no need to appeal to a soul any longer.

There have been attempts to retain the notion of a soul in a post-Darwinian world. Warren Brown, for example, attempts to understand the soul as an "emergent property of certain critical *human cognitive capacities*" (Brown, 1998, p. 103). The soul, then, is not something that is locatable in the body (i.e. Cartesian dualism) but is rather the name given to the possession of a number of inter-related capacities that together give rise to something new.

Certainly, this gets around the necessary rejection of cosmological dualism, but it ignores the fact already mentioned that there is *no* capacity that can adequately demarcate humanity from other creatures. If the soul is an emergent property, then it is not a doctrine of human uniqueness. Perhaps even more damning is that Brown concedes that, given this new understanding of the soul, "God may also relate to whom he chooses" (Brown, 1998, p. 123). In other words, if the soul is no longer a uniquely human possession, and God can relate to those without a soul anyway, it is not entirely clear why the doctrine of the soul needs to be retained as a theory of human uniqueness.

The rejection of the traditional theory of original sin

Another important theological claim for the uniqueness of humanity, albeit a negative one, is the doctrine of original sin. Because of the Fall that was initiated by the original couple, humanity is uniquely in need of salvation. Humanity is not only unique in the world: it is also unique in its relationship with God.

It has been claimed by many theologians that original sin is the only doctrine of Christian theology that is empirically true (Volz, 1992, p. 86); one can glean the truth of original sin simply by observing the world. Winter, for example, writes that

> I am convinced that the constant pattern of behavior of the vast majority of the human race ... cannot be accounted for in terms of simple psychological cause and effect ... [there is] no difficulty

> in tracing it to original sin. (Winter, 1995, p. 5; see also Haffner, 1995, p. 123)

When human relationships at every level (from international to local community to family) are examined, it is easy to sympathize with this claim. There does appear to be a lamentable tendency for humans to turn to violence and oppression as a way of dealing with their fellow human beings (not to mention humanity's treatment of non-human animals). However, despite this obvious sympathy with those sentiments, it does not follow that destructive human behaviour unmistakably points to a historical event. This connection between human behaviour and a historical event found its classic expression in Augustine.

a. Augustine

The great and supremely influential fourth-century Bishop of Hippo, Augustine, is widely regarded as being responsible for providing much of the groundwork for the classical, Western (or Latin) understanding of the Fall of humanity and original sin (Foster, 2009, p. 162). Augustine's doctrines of the Fall and sin have become "unquestionable orthodoxy" for the Western Church (Hill, 2003, p. 87). In fact, at least according to Sion Cowell, the term "original sin" was "expressly created by St. Augustine" (Cowell, 2006, p. 201). In this way, therefore, Augustine has played a crucial role in "fashioning the language" of Western theological anthropology (Mattox and Roeber, 2012, p. 72), including, importantly, both Martin Luther and Anselm (see Hinlicky, 1997, p. 41 and p. 45).

The reason for Augustine's preoccupation with original sin can be traced back to the influence of Pelagius and the heresy attributed to him. This is not to suggest that Augustine created this idea solely in response to Pelagius' claim that humanity has the ability to live a morally perfect life if it so chooses. (It should also be borne in mind, however, that all that is known about Pelagius comes from Augustine and his treatment of Pelagianism, which colours our understanding of Pelagianism, so that we always see it in the context of being in opposition to Augustine's own theology.) In fact, Pelagius emphasized humanity's ability to live a morally upright life in response to what he perceived to be a "moral laxity of Rome" due to Augustine's preaching (Passmore, 1970, p. 94). However, it is certainly within this context that Augustine cultivated and developed the concept of original sin.

The crux of the debate between Pelagianism and Augustinian "dual predestination" can be reduced to a simple disagreement. Whereas "Augustine stressed the divine initiative, Pelagius [stressed] the human response", and the theory of original sin prevailed, because the Church rightly understood that the divine initiative must also be primary (Ferguson, 1980, p. 119). Thus, Augustine maintained that if humanity could live righteously solely by its own efforts then it made "the cross of Christ to be of no effect". Moreover, if humanity could achieve eternal life on its own, then "faith in Christ was needless" (O'Grady, 1985, pp. 119–20). This led to Augustine developing the doctrine of "dual predestination", which held that those who achieved eternal life did so solely because they were chosen by God. This election to eternal life by God demanded the opposite claim that those who were destined for damnation did so because God had decreed so. Augustine reconciled this unjust claim that God condemned humanity to eternal damnation by claiming that humanity deserved it; if God saved only a few, this was out of pure goodness, and those who were not elected could not complain (O'Grady, 1985, p. 120).

Despite the fact that Augustine expounded his position in a systematic theological fashion, the underlying reason for his disagreement with Pelagius was deeply personal. It was Augustine's own struggle with sin and his own inadequacy that convinced him that humanity, no matter how hard it tried, could never live up to the moral demands of the gospel. Radcliffe makes the same point with more sympathy for Augustine's personal struggle:

> When the Pelagians insisted that all sin was a fully conscious rejection of God, Augustine replied that most sins are committed by people weeping and groaning. (Radcliffe, 2005, p. 34)

It is not enough to *want* to lead a morally upright life—humanity is simply *unable* to do so. Augustine's own weeping and groaning (influenced by Paul's assertion that "I do not understand my own actions. For I do not do what I want, but I do the very thing I hate" [Romans 7:15]) only convinced him that his theory of original sin was correct. It is not simply enough to choose how to live; humanity just could not do it, and the story of the Fall was a perfect explanation.

This personal experience, of course, does not provide proof for what Augustine argued (the human experience is varied and diverse), but it again explains why he reached the conclusions that he did. However Augustine came to this conclusion is, perhaps, unimportant. What *is* important is that what Augustine believed came to be considered orthodox in the Western Church.

b. The rejection of Augustine

Despite the fact that Augustine's doctrine of original sin has come to be accepted as orthodoxy, there are many problems with it. Perhaps the most important problem for the purposes of this book is that it requires the historical reality of the Genesis narrative, which is rejected by the *fact* of evolution, let alone the *mechanism* of neo-Darwinism. However, Augustine's theory can also be found wanting for other reasons.

b.1. Thematic issues

Augustine's particular theory of original sin can be rejected for thematic reasons, namely, a preoccupation with legal categories. This is not Augustine's fault, however, as it can be traced back to "legalism of ancient Rome" (Agourides, 1964, p. 210) and "perhaps also by Teutonic law" (Pelikan, 1971, p. 119). Theologians have always been influenced by the historical context in which they find themselves and Augustine is no different. However, this legalistic spirit can be accounted for more clearly in those who influenced Augustine. Ambrose, for example, who baptized Augustine and thus had an influence on his theological education and development, was a lawyer. Tertullian, an immensely important early theologian who came from the same city as Augustine, was also a lawyer. Winter notes that "the frequency with which [Tertullian] uses terms like purchase and ransom is an indication of the path which Latin theology would take later on" (Winter, 1995, p. 44), and, therefore, "it is interesting to observe that the forensic orientation first set by Tertullian is pursued so thoroughly by Augustine" (Winter, 1995, p. 57). "This legalistic approach to atonement," argues Winter, "was destined to be the principle orientation of Western thinking for centuries to come" (Winter, 1995, p. 58).

In other words, regardless of the Pelagian context within which Augustine argued for a doctrine of the Fall and original sin, it may have been a preoccupation with legal categories that Augustine inherited from Tertullian that fatally coloured his response.

The particular historical experience of the West could also provide another thematic reason for rejecting Augustine's theory of original sin and could also provide an explanation for a preoccupation with legal categories. Roland Chia, for example, argues that the experience of the fall of Rome forced Western theologians to "question the justice of God more than the East", which he claims influenced and shaped Latin soteriology (Chia, 2011, p. 126). Chia contends, therefore, almost in a Deuteronomistic sense (i.e. that the good are rewarded and the bad are punished [see Harris, 2003, p. 160]), that the demise and eventual collapse of the Roman Empire was instrumental in the development of the Latin Fathers' theological outlook (Chia, 2011, p. 126).

b.2. Mistranslation

Another important criticism of Augustine is his mistranslation of important biblical passages. As Augustine spoke and wrote in Latin, he therefore relied on Latin translations of the original Greek texts that made up the New Testament (and Latin translations of the Greek translations of the Hebrew Old Testament [i.e. the Septuagint]). It would be unfair, therefore, to claim that this criticism is solely down to Augustine. Nevertheless, if Augustine based his theology on inadequate translations, then this will surely impact on his theology.

The Latin translation of the Greek language scripture "was a crude mechanical translation" that did not have the same rigorous editing that the Greek Septuagint did (Mattox and Roeber, 2012, p. 72). This means that Augustine, as did all Western Christians, inherited a clumsy and inferior Bible.[13] In particular, it is the way that Augustine deals with the mistranslation of Romans 5:12. On this point Patricia Williams writes:

> The Latin mistranslation of *because* as *in whom* leads Augustine to think all people are somehow contained in Adam and carry

the guilt for Adam's sin ... [thus] Augustine's misreadings of Paul leads him to misinterpret Genesis 2 and 3. (Williams, 2001, p. 41; see also Cowell, 2006, p. 201 and Pelikan, 1971, pp. 136–7)

This particular passage is crucial for Augustine's doctrine of original sin, especially his ascription of sin to all humanity, because of Adam. Misinterpretation of this passage, therefore, undermines Augustine's rendering of the question.

b.3. Eastern Orthodoxy

It is important to note that both of these criticisms of Augustine's theory of original sin are not applicable to the Eastern Church, which did not have those problems with translation, or the same historical experience, or a figure comparable to Pelagius. Thus, another criticism of Augustine's theory is that it "was not shared by the great doctors of the Christian Orient" (Pelikan, 1971, p. 4). Augustine may have had an almost single-handed influence on the development of Western theology, but that influence did not extend into the Eastern Church. Cowell even notes that "one of the greatest minds of the middle ages", Photius the Great, condemns Augustine's belief in a "sin of nature" as heresy (Cowell, 2006, p. 201). Not only did the Eastern Church not share the Western model, at times it explicitly rejected it as heresy.

While the Eastern Church did not have a figure of comparable influence to Augustine, Irenaeus of Lyon is considered to be important for this understanding. Maximos Aghiorgoussis writes of Irenaeus:

> Unlike St. Augustine's doctrine of original justice, which attributes to the first man several excessive perfections, perfect knowledge of God and God's creation, for example, that makes the fall impossible, the doctrine of the Greek Fathers of the image of God in man as potential to be actualized, allows for the possibility of deterioration as well. St. Irenaeus speaks of the first man (Adam) as an infant (*nepios*), who had to grow up to Adulthood. Instead man failed himself, by not "passing the test" of maturity given to him by God. (Aghiorgoussis, 1992, p. 38)

The important point here is about growth. The Fall of humanity "was a necessary stage in the growth to maturity": a growth that led humanity from "infancy to maturity" (Wilken, 2003, p. 67; see also Burns, 1981, pp. 2–3). This is not a novel invention. Irenaeus is influenced by, and takes over the position of, Tatian and Theophilis, who both argue that "unfallen man was an imperfect, undeveloped, and infantile creature", which disagreed with Augustine's later claim that humanity was created in a state of "original righteousness" (Williams, 1927, p. 193). There is a significant difference in the understanding of salvation: in Latin theology it is a return to a previous perfection; in Greek it is the attainment of a perfection that has not yet been "achieved" but was the original intention of creation.

This disagreement with the Western, Augustinian emphasis on "original perfection" also meant that the Eastern Fathers were more open to an allegorical interpretation of scripture; they "knew that the account in Genesis could not be taken literally" (Wilken, 2003, p. 144). For Augustine, and the Western tradition that followed him, it was imperative that the Genesis narrative was taken literally. It was crucial to their understanding of sin that there was a literal and historical event on which to base the Fall and, importantly, the guilt of humanity.[14] N. P. Williams also notes this openness to allegory on the part of the Eastern Fathers:

> The tendency to allegorize the story of Gen 3 . . . [and] the lenient description of 'original sin', if such it can be called, as a 'slight infection', which does not seriously impair the self-determining force of man's free-will. (Williams, 1927, p. 230)

The difference between East and West, therefore, is not so much restricted to an anthropology as representing an entire approach to theology.

b.4. Textual commentary

Latin mistranslation and a tendency towards literal interpretations of the Bible are not the only biblical problems with Augustine's understanding of original sin. Modern biblical criticism also provides another important reason why Augustine's theory must be criticized. While neo-Darwinism, for example, argues that it is impossible to take a literal interpretation of the Genesis narrative, modern textual criticism argues that a literal reading of Genesis does not produce the same conclusions that Augustine came to in the first place. More specifically, modern biblical criticism claims that a close reading of the Genesis narrative cannot find any evidence of a Fall. In other words, the historical record upon which the Augustinian doctrine is based actually reveals something quite different: that "the Garden of Eden story is an etiology—that is, a myth of origins" (Kelly, 2009, p. 4).

Patricia Williams, in her book *Doing Without Adam and Eve*, observes that many biblical critics claim that Genesis 3 does not contain a Fall; nor, she continues, is it prominent in the New Testament (Williams, 2001, p. 38). Drawing on the contribution of Harold Bloom, Williams claims that the book of Genesis tells that the figures of Adam and Eve were not created perfect, and thus the punishment meted out to them for disobeying God does not "degrade or corrupt" their nature, as Augustine claims. Williams continues:

> Bloom does not find the catastrophe that Christian theologians call *the fall* in the text. On the contrary, he notes that the text explicitly says Adam and Eve become more like gods after they eat the fruit of the tree of knowledge ... the narrative never states, as later Christian authors will, that God created Adam and Eve immortal. In fact, the narrative suggests they were originally mortal [immortality is a consequence of eating the fruit]. (Williams, 2001, p. 37, p. 40)

Perhaps quite surprisingly, Williams reports that, not only is there no Fall to be found in Genesis, but also the passages that tell of the event that is often portrayed as a Fall actually tell of an "improvement" in the condition of Adam and Eve. When the couple eat the fruit of the tree of knowledge, they are actually positively transformed, as the Bible tells that they would be.

Instead, Williams argues, the Fall (as an interpretation of this passage) was the result of the early Christians' interpretation post-crucifixion as they attempted to find a reason for Christ's violent demise. Rather than understand the crucifixion within the paradigm of a fallen humanity, Williams writes that they believed that Christ "had made an atonement when he was crucified" and so they had to invent a "catastrophe" in order to explain that atonement (Williams, 2001, p. 31; see also pp. 80–1).

Interestingly, Williams continues that it is only Paul who turned to Genesis 2–3 to explain that catastrophe, whereas most of the earliest Christians turned instead to Genesis 6:1, understanding Christ as an exorcist. However, by the third century, Paul's use of Genesis 2–3 had won the day and become the dominant explanation for Christ's crucifixion (Williams, 2001, pp. 32–3).[15] It is easy to comprehend how the legalistic mindset of theologians helped to shape such an interpretation.

b.5. Evolution

All of these criticisms of Augustine's theory of original sin amount to a strong rejection of his understanding of it. Most of them centre on the fact that Augustine requires the historical accuracy of the Genesis narrative, and this has been questioned for a number of different reasons. However, the most important reason to question the historical accuracy of the Genesis narrative is the evidence of geology and evolution. Geologists such as Charles Lyell question the timescales involved, whereas biologists reject monogenism, i.e. that there is an original couple from whom the whole human race can trace its origins. While this rejection of monogenism is the result of the *fact* of evolution, the prevalence of accumulation means that it is a more marked conclusion of the *mechanism* of neo-Darwinism.

Francis Collins, a leading contributor to the project that mapped the human genome, writes that humanity must have descended from "approximately 10,000 in number, who lived about 100,000 to 150,000 years ago" (Collins, 2007, p. 126). Not only does this fit with the genetic evidence, but Collins also claims that it "fits well with the fossil record" (Collins, 2007, p. 126), further contributing to the criticism of the historical accuracy of the Genesis narrative. Ilia Delio also contributes to this rejection of monogenism for genetic reasons, by claiming that "the genetic diversity of the present human population ... could not possibly have been funnelled through a single human couple" (Delio, 2013, p. 116).

The rejection of monogenism, and the resulting questioning of the historical accuracy of the Genesis narrative, has fundamental implications for Christian theology. As TeSelle notes, it is "difficult to imagine any actual state of obedience to God, devoid of aggressiveness or lustfulness, at some stage called the beginning of the human story" (TeSelle, 1970, p. 319). While Augustine's theory of original sin may be criticized for a number of different theological and thematic reasons, it is the evidence of evolution that cements its demise.

c. Neo-Darwinian theories of original sin

Despite this rejection of Augustine's particular theory of original sin through sexual transmission of inheritable guilt from an original and historical transgression of a first couple, it does not follow that original sin itself must be completely rejected. If the openness to failure that characterizes all creaturely relationships (not just human relationships) must be retained, then a different explanation is required.

One way of dealing with the question of sin without appealing to a historical event on which to pin the transgression is to claim that it is due to something that is inherent in creatures. Teilhard de Chardin, who is perhaps the most important theologian to develop an evolutionarily sensitive theology, argued that original sin was not due to a past transgression but to a "statistical probability" inherent in an evolutionary paradigm: "evil is a by-product of evolution" (Cuénot, 1967, p. 87). He writes:

> In "classical" interpretations, suffering is *first and foremost a punishment, and expiation* ... in contrast to this view, suffering ... is primarily the consequence of a *work of development* and the price that has to be paid for it ... physical and moral evil *are produced by the process of Becoming*. (Teilhard de Chardin, 1968a, p. 71)

For Teilhard, the evolutionary process itself is one of pain and failure. The presence of evil, therefore, is not the result of a literal Fall of humanity (although Teilhard does still use this traditional rhetoric [e.g. Teilhard de Chardin, 1968a, p. 71]). Rather, the failure of humanity to live the ideal, which provoked Augustine to develop his influential theory, is the result of the ontological make-up of creation. Original sin, for Teilhard, "simply symbolizes the inevitable chance of evil" (Teilhard de Chardin, 1969, p. 40).

The implication of this realization is that original sin "is [not] bound up with human generation" (Teilhard de Chardin, 1969, p. 40), but must have a far wider application. Just as humanity cannot be treated differently from other creatures, so other creatures are affected by this openness to failure and the "chance of evil" that is original sin. Sin, understood as moral failure, is simply the human manifestation of this openness to failure that is characteristic of all creatures; as Robert North argues, "sin is evolutionary dropout transferred to the level of freedom" (North, 1968, pp. 31–5). Of course, now that freedom is not a uniquely human capacity, and other creatures are capable of displaying moral behaviour, other creatures must be capable of sinning in the traditional, theological sense of that word. Celia Deane-Drummond poses precisely this question when she asks whether dolphins could sin "inasmuch as they fail to realize their flourishing, [and become] addicted to destructive behaviour patterns" (Deane-Drummond, 2009, pp. 161–2).

However, sin, as a moral failure, is nothing more than the manifestation in relationships of a much more fundamental and primordial ontological condition of being created. Sin is simply the result of an openness to failure, which in some environments is manifested as *moral* failure. Thus, Robert Faricy, writing about Teilhard de Chardin's theory, claims:

> In non-living things, this waste takes the form of disharmony or decomposition . . . in living beings it appears as suffering and death . . . and in the moral order, in the realm of human freedom, this waste and failure takes the form of sin. (Faricy, 1981, p. 53)

Faricy, therefore, connects the physical "evils" of entropy and disharmony and the natural evils of suffering and death with the moral evils of sin. Such a connection is not possible without a very definite anthropology based on the connection of humanity with the wider world.

Despite the fact that this is a nuanced interpretation of original sin—which moves it away from a specifically human problem based on a historical transgression to seeing it as a much wider condition of being created, and conducive to a neo-Darwinian context—it is not a novel view. This connection of human sin to the wider, non-human sphere has Patristic precedence. N. P. Williams, for example, notes that this idea can be found in the theology of Gregory of Nyssa:

> Gregory seems to recognize, in a startlingly modern spirit, that all so-called 'evil passions' in man are exaggerations, or perversions, of instincts which in themselves are necessary for the continuance of animal life upon earth. (Williams, 1927, p. 272)

The natural instincts that, when "exaggerated or perverted", are responsible for sin cannot be sinful in and of themselves "because they lead us to garner resources so we can live and reproduce" (Williams, 2001, p. 151; see also Hefner, 1993, p. 139). In *The Life of Moses* Gregory of Nyssa elaborates that too little of a virtue produces sin, as does too much of it, but in the "right amount" a particular virtue is good (Gregory of Nyssa, 1978, p. 128). Sin is not a uniquely human phenomenon. Rather, sin must be considered a *creaturely* phenomenon because (a) non-human creatures can display moral behaviour, and (b) sin is essentially an openness to failure that characterizes all creatures (as "disharmony", "death", or moral failure), precisely because there is no ontological distinction between all creatures.

—

There is a connection here between original sin as a permanent ontological condition of failure that characterizes what it means to be created, and what was concluded about neo-Darwinism in the previous chapter. Original sin is not a temporary, unnatural phenomenon from which humanity requires salvation; it is a permanent and natural phenomenon that simply characterizes and describes what it means to be a creature. Human sin is no longer infused with guilt; failure is not the result of a specifically human, historical misdeed, and therefore is not something from which humanity, or any other creature, requires salvation. This does not mean that the failure that characterizes human relationships should be accepted, but it does mean that it is impossible to be created without this openness to failure. Thus, "what is not God will by definition be subject to imperfection, decay, collision, conflict" (Shortt, 2016, p. 88).

Evolution resists the failure that produces change, but it is unable to resist perfectly. The same conclusion must be applied to humanity (after all, genes are just as much creatures as humans). Humans must resist the failure that separates them from God, but this failure is not something from which creatures need salvation; not being God (i.e. open to mutability) is an ontological condition, but not a negative one.

Neo-Darwinism, then, does not deny original sin, or at least it does not deny the content of that doctrine, namely, that creatures are open to failure as their very nature. What neo-Darwinism does deny is that this condition is a temporary one that is wrought on creation as punishment for the fault of an original human couple. It therefore denies also that this condition is something from which creatures either need saving or can be saved. Neo-Darwinism, along with biblical criticism, denies the historical underpinning of original sin. This serves to deny that humanity is the subject of a catastrophic Fall—indeed, modern biblical criticism cannot find a Fall to begin with. In other words, the condition of humanity (and the rest of creation, from which humanity can no longer be considered apart) is not a punishment. In the same way that Gould distinguished between "fact" and the "mechanism" of evolution, so, Ronald Cole-Turner argues, it is important to distinguish between the "affirmation" of disorder and its "explanation"; the disorder must still be affirmed but the explanation must be denied (Cole-Turner, 1993, p. 84).

Drawing on the openness to allegory that characterizes the Eastern idea of original sin, this means that the Genesis narrative does not have to be completely rejected. It can still be an allegorical myth describing what it means to be created, but this is not the result of a punishment meted out on the basis of the misbehaviour of primitive, perfect ancestors. That condition, outlined by an allegorical interpretation, is that humanity (and creation) are open to the possibility of failure; this is nothing but the conclusion of neo-Darwinism.

The Genesis narrative is an etymological myth that seeks to explain the human condition of failure through story and narrative; the historical accuracy of the story can be rejected, but this is still the point behind it. Neo-Darwinism is a modern biological and scientific theory that explains this very same phenomenon (although that was not the intention behind the formulation of the theory). Darwinism agrees with the human condition that the Genesis authors wanted to explain—it simply offers an explanation using different tools. It is important to make clear that this does not mean there is a causative relationship between original sin and evolution—original sin is not *responsible* for evolution, nor vice versa. It simply means that neo-Darwinism and Christian theology agree on the question of ontology, i.e. what it means to be created.

Yet it is important not to overstate this point. To say that neo-Darwinism and the Genesis narrative reach the same conclusion regarding the place of original sin (i.e. that creation is open to failure) is not to claim that science supports it. Some dismiss scientific advancement as merely confirming what the Bible and religion have been claiming for millennia. Yet, such concordism fails to do justice to the subtleties and nuances of both positions (see Cuénot, 1967, pp. 47–8). Further, such concordism fails to acknowledge, as this and the next chapter illustrate, that neo-Darwinism (and modern science) contribute much to a *reinterpretation* or *nuancing* of anthropology and theology. (This concordism does not even take into account that "what modern readers might think of as the literal reading of a particular passage, based on a superficial reading, might not at all correspond to what the ancient author might have meant to convey" [Oakes, 2016, pp. 98–9].)

Neo-Darwinism affirms the core message of the Genesis narrative (namely, that creatures are open to failure), but it crucially develops a nuanced and different understanding of this core. To claim that neo-Darwinism and Genesis *agree* is not to say that they are *identical*.

It should be made clear and explicit that the *truth* that the doctrine of original sin seeks to formulate is not denied here—indeed, it is upheld. Rather, the language and imagery used in its traditional formulation is claimed to be outdated to the point of incoherence. The traditional notion of original sin, as evil in a "fixed world" of "original perfection" (Delio, 2013, p. 116), no longer makes any sense in the modern world. Neo-Darwinism seeks to explain the same reality, yet using different language.

However, despite the fact that Augustine's theory has been rejected, this does not mean the Church was incorrect in siding with Augustine over Pelagius. What separates the theories of Augustine and Pelagius is what was responsible for sin; for Pelagius, who holds that humanity can be morally upstanding if it so chooses, sin is a matter of the will, whereas for Augustine sin is a matter of nature. Admittedly, for Augustine it is unnatural in the sense that it is not the desired state, which neo-Darwinism disagrees with, but the essential point is that humanity cannot choose to be perfect—it is part of its ontology.

In this way, neo-Darwinism supports the idea of original sin; it simply adds that, because this is a permanent condition of being a creature, it is not something from which creatures can be saved. Secondly, the tension between Augustine and Pelagius was a concern about creatures' need for and complete dependence on divine grace. If sin is about creatures' will, then they are no longer completely dependent on divine grace: they can do good by themselves; they simply need to choose to do so. However, if sin is about creatures' nature, then it is no longer a question of choice: creatures are dependent on divine grace regardless of their moral aptitude. In this way Augustine must be correct. Fergus Kerr observes:

> The doctrine of original sin encapsulates an idea of human inadequacy because of human dependence on divine grace, which amounts to a denial of human self-sufficiency. (Kerr, 1997, p. 129)

The question of original sin is about human dependency on God; in this way, Augustine must always be correct, and the Church was always correct to side with him.

Conclusion

The neo-Darwinian synthesis has a profound effect on theological anthropology, but this does not mean that it is changed out of all recognition. Rather, it could be seen to provide a positive interpretation of the essentially negative anthropology that theologians, perhaps wrongly, glean from the Bible. This chapter has, therefore, made two adjustments to the way that theology understands the role of humanity in the universe and its relationship with other creatures and God.

In the first place, by rejecting the historical accuracy of the Genesis accounts of creation, the primacy that humanity has traditionally held is rejected. Much more important than the rejection of human primacy, however, is that the subjective nature of value judgement and the prominence of accumulation mean that the ontological separation of humanity from the rest of creation is no longer possible. There is a natural, genealogical, and familial unity between all creatures, both spatially and temporally. There is nothing that can separate humanity from the rest of creation.

The uniqueness of humanity is also rejected when it comes to the doctrine of original sin. The rejection of the historical Fall means that the condition of openness to failure that is the content of original sin is not a temporary phenomenon that is the result of a historical human transgression. Instead, sin is a permanent and natural condition of what it means to be created. This has a number of implications. Firstly, it means that humanity is no different from the rest of creation: all creatures experience, and live in a state of, an openness to failure. Secondly, it means that there is a close connection between what the Genesis author(s) were trying to communicate and the conclusions of the neo-Darwinian synthesis: to be created is to be open to failure.

This means that the failure of genes and the failure of humanity are essentially manifestations of the same phenomenon. This likewise means that sin can no longer be viewed as a negative doctrine, understood as a punishment meted out for disobedience and requiring salvation. Just as neo-Darwinism provokes a more positive interpretation of the relationship between humanity and non-human creatures, so it provokes a more positive interpretation of the relationship of creatures to God.

This does not represent a fundamental re-casting of theological anthropology. Neo-Darwinism agrees with theological anthropology much more than it disagrees with it. However, what it does do is widen the scope of what theological anthropology claims. Neo-Darwinism is a far more inclusive and positive rethinking of what theological anthropology already claims about the place of humanity in the universe.

The question of neo-Darwinian implications for divine activity will be explored in the next chapter; however, it is important to anticipate that discussion by briefly noting how this affirmation of an inherent unity of all creatures agrees with, and provides further evidence for, what will be concluded. If there is an inherent natural and genealogical unity between all creatures, such that it becomes impossible to satisfactorily demarcate and separate humanity from non-human creatures, then humanity can no longer claim a monopoly on divine influence: what God does for humanity, God does for all creatures. Hefner points to this conclusion, claiming:

> The circumstances of human creation and development would seem to imply that God's will for us humans transpires within the larger realm of the divine will for the entire natural order, as creation. (Hefner, 1993, p. 60)

God's will for humanity, writes Hefner, can only be understood as part of a wider will for all creatures. Humanity no longer monopolizes the will and energy of God. Humanity is no longer favoured by God. God is no longer uniquely concerned with humanity and the human story; instead, there is a unity of divine activity. God does not do one thing for humanity and another for non-human creatures; God treats all creatures equally. Anticipating the conclusion of the next chapter, this means that God does one thing; the unity of divine activity does not mean that God does different things that have a unity in intention, but it means that God literally does the same thing for all creatures.

CHAPTER 3

Divine Activity

In Chapter 1, "The Neo-Darwinian Synthesis", it was argued that several important pieces of evidence necessitate the conclusion that God does not control evolution. Despite the fact that the expansion of the scope of neo-Darwinism beyond biology has been criticized by some, many have noted that the simple mechanism that Darwin and his successors explained has a universal application; as Dawkins explained, the gene is just one replicator among many. This widening of the scope of the neo-Darwinian synthesis means that the conclusion that God does not control evolution must be widened to include the whole universe. If God does not control evolution, it is because God does not control/influence *anything*. The neo-Darwinian synthesis, therefore, makes very particular suggestions for what it means to talk about divine activity; this chapter will explore these.

Throughout the history of theology there have been many attempts to explain how God influences the world. One of the most enduring is to claim that God uses the scientific laws of the universe to control the world (see Augustine, 1991a, p. 130). This assumption lies behind the claim made by many theologians that God uses evolution to create for him. One interesting modern version of this theory is Arthur Peacocke's "whole-part" influence, in which he argues that God influences the world as a spatially and temporally complete whole, with divine influence "trickling down" to influence specific events (Peacocke, 2001, p. 110). If the evidence of neo-Darwinism is to be taken seriously, however, then this can no longer be accepted as a viable theory of divine influence. Despite the fact that Peacocke claims that God would not be "intervening within the supposed gaps" (Peacocke, 2001, p. 109), he still assumes that God can influence the world.

In the twentieth century, after the scientific method had become so adept at explaining how the world works (and, thus, closing the gaps in which God can "intervene"), both chaos theory and quantum mechanics were considered as possible theories of divine influence. Theologians who look to chaos theory as a possible place of divine activity argue that "an infinite level of accuracy in the initial positions of [what is being calculated] is required to make an accurate prediction" (Saunders, 2002, p. 180). This means that, in practice at least, there is an indeterminacy in the world that God could influence without science being able to detect it. However, that indeterminacy is not "built into" the fabric of nature; it is a theory of infinite measurement rather than inherent indeterminacy.

Quantum mechanics, on the other hand, has an advantage as a theory of divine activity because it claims that quantum interactions are indeterminate in principle (Russell, 2008, pp. 21–2). In other words, quantum mechanics argues that the gaps in scientific knowledge are not due to a lack of precision in measurement: they are genuinely part of the world. In this way, the scientist can never claim that an indeterminate event was not determined by God, because it is indeterminate in principle.

However, much like the criticism of Peacocke and secondary causality above, these theories still assume that God *can* influence creation. They do not question the ability of God to influence the world; they simply look for a gap of indeterminacy in which to hide divine activity. Rather than taking the evidence of scientists and biologists at face value, they try to find a way around the evidence, more often than not by belittling science itself and claiming that it is incapable of recognizing divine influence, or that it can never fully explain the world.

—

This chapter, taking its lead from the conclusion of Chapter 1, that God does not influence evolution, will show how the neo-Darwinian synthesis, taken in conjunction with other scientific advancements (particularly Einstein's equation of matter and energy, and the implication of his theory of relativity that space and time form a four-dimensional whole, rather than a separation of space and time), demands the conclusion that God cannot influence the world. Moreover, it will show that this rejection of divine influence can find support in traditional understandings of the eternity and transcendence of God. Neo-Darwinism, therefore, by disagreeing that God can influence the world, does not reject the idea of God; rather, it translates the traditional classical understanding of God into modern scientific language.

This discussion will take place through three sections, each with a different aspect of divine eternity and showing how these need to be nuanced to describe a God that can agree with the conclusion that neo-Darwinism has offered. If Einstein is correct that time and space are the "same" (i.e. they are both dimensions), such that temporal extension in one dimension is the same as spatial extension in another (i.e. the spatial indexical "here" is entirely parallel to the temporal indexical "now" [Craig, 2001, p. 127]), then the eternity of God has just as many spatial connotations as it does temporal. If God is non-temporal, then God cannot have any spatial dimensions either, both of which strongly imply that God cannot cause any material effects.[16]

The first section of this chapter will argue that the non-spatial dimensions of divine activity mean that it makes no sense to claim that God can influence particular events, as this would mean being able to locate divine activity (i.e. claim it is "here" rather than "there"). It will also be shown that this is nothing more than a translation into modern scientific language of traditional understandings of the transcendence of God, i.e. divine activity is of a different "kind" than created activity.[17]

The second section will argue for the same conclusions but from the non-temporal dimensions of divine activity. It will be argued that being unable to apply temporal language to divine activity not only means that it cannot be located temporally (i.e. claim that it is "now" rather than "then") but also that succession and reaction must also be rejected. This means that God only does one thing once; yet that one act cannot be located temporally. God acts eternally. This leads to the conclusion that God does not really *act* at all; rather, divine activity is a *relationship* between God and creatures—a relationship that creates and sustains.

These considerations lead to the third section of the chapter: the claim that divine activity must be simple. God only does one thing, and God only acts once. Just as God cannot be divided either spatially or temporally (i.e. God does not have "parts" or "succession"), so divine activity must likewise be indivisible. This means that God only does one thing once, and that one thing is the conservation of the universe, i.e. keeping it in being.

The eternity of God: divine immateriality

In Chapter 1 it was claimed that theologians who argue that neo-Darwinism must always remain neutral on the subject of God's control of evolution, because science cannot detect that control, were incorrect. The four main points that were made—that natural selection is not a real phenomenon (i.e. it is an observation of the differential survival of individuals); the primacy of preservation (i.e. without accumulation, change cannot have an effect on survival); that both genes and natural selection *resist* that change; and that change (the only raw material of selection) is the result *only* of accidental malfunction of genetic replication—all provide evidence to the contrary. Neo-Darwinism does not remain neutral about God's involvement; it argues against the idea that God controls evolution. There is not only an *absence* of evidence of God's influence—there is *evidence* of absence of God's influence.

Such a claim also forms one of the crucial arguments of Richard Dawkins' famous *The God Delusion*. Here he argues that, regardless of how much one claims that the scientific method cannot comment on theological questions, if theologians claim that God influences the world (which *is* the domain of the scientific method) then scientists *can* make comments on theological matters (Dawkins, 2006a, p. 184). If theologians argue that God influences the world, then there will be points at which the scientific method can comment, and if the scientific method cannot find any evidence, then this must be taken seriously. Other commentators agree. Nicholas Saunders writes that if theologians claim that divine activity "does achieve causal physical effects", then those effects would "surely be epistemologically open to scientific analysis" (Saunders, 2002, p. 129). Frank Kirkpatrick, too, says:

> A God who acts and whose acts decisively change the course of history forever but who can't be known through those acts is

> a peculiar God and so is an epistemological situation in which knowledge of God through God's own acts is impossible or at least profoundly difficult. (Kirkpatrick, 2014, pp. 136–7)

Again, if God could influence the world then those points where God does influence would be open to scrutiny.

However, Dawkins makes one logical leap too many. Whilst he claims that this proves God does not exist, what it actually proves is that God does not influence the world. Theologians are correct that the scientific method cannot make comments about God's *existence*, but it can make comments about God's *activity*, if that activity is claimed to directly influence the world. The lack of evidence of divine activity, not to mention the evidence that neo-Darwinism offers *against* this proposition, strongly suggests that God does not influence the universe, not that God does not exist.

—

It is not just the evidence of neo-Darwinism that strikes a blow for divine activity. Philosophical claims about the ability of the immaterial to influence the material also serve to reject the claim that God can influence the world, even if God so desired. As an argument against cosmological dualism, particularly in relation to the brain–mind problem, many philosophers and psychologists argue that something that is immaterial, i.e. the mind, cannot influence something that is material, i.e. the brain, thus rejecting the idea that the mind is separate from, and uncaused by, brain states. For consciousness to be able to influence the body, it must be completely explainable through materialistic means. This is termed the "interaction problem", of which the mind–brain problem is just one example and divine activity must be another.

Further evidence for the so-called "interaction problem" and its disagreement that the immaterial can influence the material is the conservation of energy. This is the scientific principle that there is a finite and set amount of energy/matter in the universe that can neither be increased nor diminished. For the immaterial to influence the material, it is argued, there would have to be the appearance of new energy, thus violating this principle. Owen Flanagan, for example, writes:

> If we accept the principle of conservation of energy, we seem committed either to denying that the nonphysical mind exists, or to denying that it could cause anything to happen, or to making some very implausible ad hoc adjustments in our physics. (Flanagan, 1991, p. 21)

Just as it denies the presence of a non-physical mind, so the conservation of energy rejects the idea that God can influence the universe. More specifically, it argues that if God wants to influence the world, then God has to be material—a claim that would contradict the idea that God is the creator (i.e. if God creates matter then God cannot be material).

It has been suggested that it is possible to get around the problem of the conservation of energy by claiming that God's activity does not move creatures by expending/introducing energy, but through a "pure entry of information" (Fiddes, 2000, p. 146). However, critics of this suggestion point out that "while information flow is not the same as flow of matter or energy ... no information flows without some exchange of energy" (Murphy, 2008, p. 129). If God is immaterial, then God cannot influence the universe.

In a similar direction (and drawing on the mind–brain problem), John Polkinghorne argues that if God can "interact with the depths of our psyches" (i.e. God can influence our thoughts), and our psyche is a "material process of our brains" (i.e. there is no immaterial mind), then God must be able to influence other matter as well (Polkinghorne, 1989, p. 10; see also Tracy, 2008, p. 276). However, as it has just been argued, there cannot be a flow of information without a flow of energy. Polkinghorne assumes that God can influence the mind and extrapolates that God must be able to influence matter; this chapter has argued that the conservation of energy rejects the suggestion that God can influence matter and extrapolates that God cannot influence the mind/consciousness.

The discovery of the neutrino can provide another example of what is meant by the interaction problem and its application to God. Neutrinos are virtually undetectable because they interact with hardly anything. Cox and Forshaw write:

> Neutrinos are ghostly particles that hardly ever interact with anything, and as such, most of them stream out from the sun as soon as they are produced without hindrance . . . the neutrinos nearly always pass through our hands, and in fact the entire earth, as if they did not exist. However, on rare occasions, a neutrino will interact, and the trick is to build experiments that are able to catch these extremely rare events. (Cox and Forshaw, 2009, pp. 163–4; see also p. 181)

There is a correlation, therefore, between detectability and influenceability or interaction; the more interactions that particles make, the more detectable they are. This means that, as Cox and Forshaw write, a particular experiment was needed, which required complex machinery, in order to detect the neutrino. They continue that "as a result, the experiment is able to 'see' the neutrinos streaming from the sun" (Cox and Forshaw, 2009, p. 164).

The implication here is that were the neutrino to fail to interact with anything then it could never have been detected. This would not, in fact, lead to the conclusion that it did not exist; yet, from an operational point of view, it would be completely functionally redundant and, therefore, unnecessary to describe the working of the universe. There would be no way of determining its existence one way or another, but it could be said with certainty that neutrinos have no influence on the universe. As Dawkins wrongly claims then, the absence of evidence of divine influence in the universe does not prove that God does not exist, only that God "has no discernible 'utility function'" in the universe (McGrath, 2005, p. 57). Vogel writes:

> Even if the existence of God is not necessary to complete a philosophy of nature . . . God may in fact happen to exist. Not everything that is true needs to be true in order to preserve the world's intelligibility. (Vogel, 1996, p. 21)

The interaction problem does not disprove the existence of God, but it does provide strong evidence that the theologian should not expect God to be able to influence the universe. As above, if God influences the world, there would be instances where the scientific method can comment; at these proposed instances (such as evolution) not only is there an absence of evidence but there is also evidence of absence. However, taking into consideration the philosophical and scientific problems, such as interaction and conservation of energy, it should not even be *expected* that God *could* influence the world.

a. Theological agreement

Despite the apparent audacity of the claim that divine influence on the world must be rejected, this is not altogether unfamiliar to theology. The distinction between primary and secondary causality that characterizes many theologians' theology of divine activity is first and foremost a claim that divine activity is fundamentally different from created or material cause and effect. To call God the primary cause is not to claim that God precedes created causes temporally—it is not a "first" cause—but to claim that God's activity is completely different.

The point about the distinction between primary and secondary causality, therefore, is not a temporal distinction but an ontological one. God's activity (i.e. primary causality) does not precede created activity (i.e. secondary causality), as if what God does and what creatures do is essentially the same *kind* of activity, but only simply temporally differentiated (i.e. divine activity happens first—creation—and then creaturely activity happens after). Rather, God's activity is a *different kind* of activity; or, as some theologians put it, divine activity and creaturely activity are not in competition with each other (see Kirkpatrick, 2014, p. 115 and Wegter-McNelly, 2008, p. 307).

Gerard Verschuuren makes this point in his book *Aquinas and Modern Science*. He writes that "when Aquinas speaks of a 'first' cause, his concern is logical hierarchy rather than temporal priority", so "creation must come first in the order of primacy, but not in the order of time" (Verschuuren, 2016, p. 47, p. 100). The difference between divine and created action is such that they "operate" on entirely different levels and, furthermore, to claim that God interacts with the universe "degrades the primary cause to a secondary cause" (Verschuuren, 2016, p. 169). Indeed, attempts to find divine activity in the gaps in quantum mechanics are exactly this, a degradation of divine activity to secondary causality (i.e. it makes divine activity the same *kind* of activity as creaturely activity). Thus, divine creating is not *post nihilum* but *ex nihilo*. God does not act *before* nothing but *from* nothing (Verschuuren, 2016, pp. 102–3).

On this reading, it would be entirely appropriate to claim, on the basis of Thomistic theology alone, that God cannot interact with the universe because divine activity is not the sort of activity that could influence the universe. This distinction is so important that it leads to the conclusion that divine activity cannot be *efficient* in any way. As Verschuuren continues:

> God creates a universe in which things have their own causal agency, their own self-sufficiency... God does not act as part of a process, nor does God initiate a process where there was none before. (Verschuuren, 2016, p. 104)

God does not initiate the universe by "pushing the first domino", because this "pushing" is a created activity, i.e. an act that influences matter. Rather, God's act grounds created activity. God does not cause the Big Bang; God simply *forms* an ontological condition in which instability is possible (the same ontology that is supported by neo-Darwinism and original sin) and the universe can spontaneously create itself. God does not need to "light the blue touch paper" to start the universe; rather, "because there is a law like gravity, the universe can and will create itself from nothing" (Hawking, 2010, p. 180).

Verschuuren summarizes this debate by claiming that "what Aquinas does in all of this is safeguard God's *transcendence*" (Verschuuren, 2016, p. 48). This is important. What Aquinas does by differentiating between primary and secondary causality is exactly what neo-Darwinism (and science in general) does: it argues for the transcendence of God. It may be too much to suggest that neo-Darwinism directly supports Thomism here, but it certainly does not disagree with it.

This is quite an astonishing claim. If God is understood in the Thomistic sense, then there is a coherence between modern science (which is impossible without, and fundamentally shaped by, evolution [see Birx, 1991, p. 113], and by extension neo-Darwinism) and theologies of divine activity as primary cause. Not only is there evidence against divine control of evolution, the theologian should not *expect* God to control evolution, because controlling evolution would be a secondary act, which would mean God would have to be material. Thus, Frank Kirkpatrick claims that "God cannot be the kind of reality that can *literally do* anything at all" because "doing" is not the "kind of reality" that God is (Kirkpatrick, 2014, p. 1).

b. The equivocality of being
The conclusions so far reached can be explained differently using the Scholastic language of univocality and equivocality of being. The medieval Franciscans, Duns Scotus and William of Occam, both argued that it was possible to have categories, such as "being", that could be applied equally to both creatures and God. However, as Robert Barron notes, this means that God, regardless of how almighty and all-knowing God is, is "one reality alongside others and hence competitive with them" (Barron, 2015, pp. 27–8). There can be nothing that creatures and God have in common.

"A consequence of this conception", writes Barron in an earlier work, "is that God and finite things have to be rivals, since their individualities are contrastive and mutually exclusive" (Barron, 2007, p. 14). This would mean that God would be in competition with creation. It would mean that divine activity is exactly the same as created activity and, by extension, would imply that God is material. Langdon Gilkey agrees with Barron, saying:

> Put in the language of contemporary semantic discussion, both the biblical and the orthodox understanding of theological language was univocal. That is, when God was said to have "acted", it was believed that he had performed an observable act in space and time so that he functioned as does any secondary cause; and when he was said to have "spoken", it was believed that an audible voice was heard by the person addressed. In other words, the words "act" and "speak" were used in the same sense of God

as of men. We deny this univocal understanding of theological words. (Gilkey, 1983, p. 32)

If God, as neo-Darwinism implies, cannot act in the world, then it must be because God and creatures are equivocal. Even if God wanted to, God could not influence the world; this includes, as it must, the claim that God "started" the universe, since "starting" the world has a material effect and thus must be "secondary".

Of course, it must be noted that the rejection of the univocality of being did not result in an acceptance of the equivocality of being for the Scholastics, such as Aquinas. Instead, the Scholastic theologians argued for a middle position, namely, the analogy of being. God could not be understood in the same way that creatures are (i.e. univocally)—this would make God a creature—but God could not be completely unknowable (i.e. equivocal) since God was responsible for creation, and creation reflected divinity. Humanity was created in the image of God, so there must be *some* sort of relationship between creation and God.

The doctrine of analogy is treated in the next chapter, but for now, it is important to maintain that God and creatures are completely differently different and "otherly other" (Barron, 2015, p. 8). So differently different that God and creatures cannot even be opposites (Shortt, 2016, p. 8), because a comparison can only be made between two things that have "some common respect" (Turner, 1998, p. 42); to be able to make such a comparison, God and creatures would have to be univocal. If God and creatures are not univocal, then they are equivocal. God is completely and utterly different to creatures and his action is of a completely different kind to created action. Divine activity does not compete with, and therefore does not influence, created activity. It makes no sense, therefore, to claim that God can influence creatures.

—

However, this also has other implications that are important to note here. It means that Boethius' definition of eternity as "the total and perfect possession of life without end" (Boethius, 2000, p. 110) must be incorrect. Eternity is not the possession of all time and space but the complete *absence* of time and space, because "eternity is neither time nor part of time" (Gregory of Nazianzus, 1894, p. 347). Aquinas agrees that eternity is "not a very long time, nor an endless infinite time. It is no time at all" (McCabe, 2016, p. 110). If eternity, which describes God's being, was the full possession of life, or the immediate full possession of all time, then this would mean that God and creatures are univocal, because there would be a category (either life or time) that is applicable to both.

Furthermore, this same logic also means that the basic attributes that are often applied to God must also be rejected. The omnipotence of God has already been rejected on the basis that to claim that God is all-powerful is simply to claim that God has an infinite amount of something that creatures have—i.e. that divine power is simply an infinite form of creaturely action—which would make God and creatures univocal. Likewise, therefore, the omniscience of God must be rejected on the same basis. Not only does the claim that God can have knowledge imply that God has a brain (as there is no immaterial mind/consciousness that is not causally related to a physical brain); it also implies that God simply has an infinite degree of the *same* knowledge that creatures can have, which again would make God and creatures univocal. God, like the Parmenidean being, "neither sees, nor hears, nor thinks, nor acts" (Passmore, 1970, p. 33).[18]

To put this into different language, God is *nothing*; as Kirkpatrick writes, God is "transcendent to the point of absence" (Kirkpatrick, 2014, p. 42). To claim that God is nothing does not make a value judgement on God, or God's worth, nor does it point to atheism, but it simply accepts that God is so equivocal that God cannot be described by anything and cannot influence anything. Again, this is the view of many theologians. Denys Turner, for example, notes that "in so far as we consider the existence of creatures, it is better to say that God is 'nothing', because God is not any kind of thing" (Turner, 1998, p. 164). Montagnes, too, writes that "if the realities of this world qualify as being, then God is not being, but beyond being" (Montagnes, 2004, p. 66). However, it is most notable in "negative theology". Pseudo-Dionysius, in arguably one of the most influential theological treatises of all time, *The Mystical Theology*, writes:

> [God] is not a material body, and hence has neither shape nor form, quality, quantity, or weight. [God] is not in any place and can neither be seen nor be touched. [God] is neither perceived nor is [God] perceivable. [God] suffers neither disorder nor disturbance and is overwhelmed by no earthly passion . . . [God] passes through no change, decay, division, loss, no ebb and flow, nothing of which the senses may be aware. (Pseudo-Dionysius, 1987, pp. 140–1)

God, continues Pseudo-Dionysius, 'is completely unknown and non-existent' (Pseudo-Dionysius, 1987, p. 263). There is nothing in this passage that a neo-Darwinian theologian would not argue for; the neo-Darwinist would add, however, that this shows that God cannot influence the world.

The eternity of God: the non-temporality of God

If Einstein is correct that time and space are inter-related (such that time and space are more or less the same "thing", i.e. a dimension of matter/energy), then the claim that God is immaterial must have as many temporal implications as it does spatial. If God cannot influence the world because God is not material, creating problems for interaction and conservation of energy, then God cannot influence the universe because God is not temporal.

The rejection of the application of temporal categories to God is not a novel suggestion. Perhaps the most famous treatment of the inappropriateness of applying temporal categories to divine activity can be found in Augustine's *Confessions*. In the chapter dealing with the problem of time, he writes:

> Since, therefore, you are the cause of all times, if any time existed before you made heaven and earth, how can anyone say that you

> abstained from working? You have made time itself. Time could not elapse before you made time. But if time did not exist before heaven and earth, why do people ask what you were then doing? There was no "then" when there was no time. (Augustine, 1991, pp. 229–30)

The immediate context of this quotation is the question of what God was doing before God created the world, to which Augustine replies that this question does not make any sense, because there was no time before the creation of time for God to be doing anything "before" there was time.

What is important about Augustine's questioning of time and its application to God is that he has been vindicated by modern physics. Paul Davies writes:

> The idea that time does not stretch back for all eternity but was created with the universe was anticipated in the fifth century by St. Augustine. There is thus a scientific counterpart to the creation *ex nihilo* of the Christian tradition. (Davies, 1989, p. ix)

This is not to suggest that modern science argues *for* a doctrine of a divine creator—it merely suggests that Augustine was not incorrect in arguing for a God that cannot be described by temporal language.

The question of the temporality of God has already been touched upon in relation to the idea that, according to Aquinas, primary causality does not *temporally* precede secondary causality. One of the reasons for this is that temporal language cannot be applied to God. If, for example, spatial language cannot be applied to God, so that God cannot be "up there", then the same must be said for temporal language. If time and space are the same "thing", such that the words "here" and "there" have the same implications as "now" and "then" (see Craig, 2001, p. 127), then language such as "before" or "after" cannot be applied to God. Or, to put this same point differently: if God cannot be located spatially, then equally God cannot be located temporally. To claim that God is eternal is not to say that God is ever-present; eternity is no time (and no space), and so this means that God cannot be present temporally (just as God cannot be present spatially). God is not present at all times and all places; God—because eternal—is not present at all.

This restriction on applying temporal language to God must be equally relevant to *all* temporal language, including temporal adverbs that are used to describe divine activity (e.g. "always", "never", etc.). A popular way of characterizing theism, particularly in an evolutionary context where evolution is deemed to be the way that God creates, is to claim that God *continues* to create, or that divine activity is a *creatio continua*. This means that God's act of creation is not limited to a point at the beginning of time, but continues and perdures through time, constantly maintaining and bringing new things into being. As Sammeli Juntunen puts it:

> All conservation is actually continuous creation. The *creatio ex nihilo* continues through every moment of existence of the created individual. (Juntunen, 1998, p. 138)

God does not create, then stop, then do something else (e.g. save); the divine influence persists throughout creation.

However, as much as the sentiments expressed here are noble, even this must be criticized. God cannot *continue* to do anything, because *continue* is so inherently temporal. God does not continue to create, because this would imply that God is subject to the succession that characterizes time. This does not mean that the deist is correct that God only acts at the beginning, because this too seeks to locate divine activity temporally. Both ideas are equally incorrect because both apply temporal categories to God.

To claim that God continually creates is no different from the deist conception, except that it does not claim that divine activity can only be located at the beginning of time, but that divine activity can be located at many different points throughout history. Rather, in opposition to both of these claims, God acts *eternally*; this means that divine activity does not have any temporal or spatial location or point—"eternity is not before, during, or after anything" (Conee and Sider, 2005, p. 40). Divine activity is of a completely different kind. In this direction, Jack Mahoney writes:

> The chain of temporal events in world and human history is integral to our existence as physical creatures that operate in a context of before and after. But in some utterly mysterious way, such events are all simultaneously in the present to God in whom there is no succession, no before, and no after. (Mahoney, 2011, p. 104)

If there cannot be any succession in God, then God cannot influence the world, because influence implies succession (i.e. a before and after [even if it is a continuous before and after] or a cause and effect). If God caused a material effect, this would be a temporal location as much as a spatial location; both must be rejected. It would be the same as claiming that God did create "then" and God does not create "now"; divine activity would be temporally locatable. As it was argued in relation to Dawkins above, once you claim that God influences the world then there must be a point that is open to scientific investigation; God would be spatially and temporally locatable.

The rejection of succession or continuation in divine activity means that God's activity does not have a beginning, nor does it have an end; it simply eternally happens. God does not start to act, nor does God finish acting; God simply, eternally, acts once without beginning, continuation or end. However, importantly, claiming that divine activity does not have a beginning does not mean that creation is eternal. As Craig writes, "God need not begin to do anything, then, in order to create a world with a beginning" (Craig, 2001, p. 213). Claiming that divine activity does not have a beginning does not imply that the universe does not.

The reason for such mutual exclusivity between the beginning of the universe and the lack of beginning of divine activity is possible precisely because eternity is not infinite time, or the "simultaneous possession of all time" but the complete absence of time. To claim that there is an identity between the eternity of divine activity and the beginning of the universe would be to make God and creatures univocal. The mutual exclusivity between the beginning of creation and the lack of beginning of divine activity is also explained, therefore, because divine activity is primary, not secondary.

It has already been claimed that to try to find divine activity in the gaps in scientific knowledge (even "permanent" epistemological gaps such as are present in quantum mechanics) would be to relegate divine activity to secondary causality (i.e. material cause), and the same applies here. To claim that there is a connection between the beginning of time and the lack of beginning to divine activity (i.e. to make divine activity directly the cause of the Big Bang) would be to make divine activity secondary, rather than primary. As Craig writes, "creation is concerned with ontological origin, not temporal beginning" (Craig, 2001, pp. 211–12). God eternally defines what creation *is*, "after" which the universe can create itself from nothing. God's act of creation is to account for why it is that the universe is unstable (i.e. the neo-Darwinian synthesis) and everything else can happen by itself.

―

The rejection of temporal succession in divine activity has already been hinted at in Chapter 1, in the rejection of the tension between finished and unfinished creation. Evolution does not bring about the completion of the universe, because the ontological condition that neo-Darwinism describes is permanent. This same point was put differently by claiming that evolution is *teleologically neutral*. Of course, evolution does not represent divine activity, but this still corresponds to the fact that divine activity does not have an end. The point here is that it does not make sense for the universe to be finished. There is nothing teleological about the universe. Even if God *did* control the outcome of evolution, it would not be teleological (i.e. have an end in sight).

Therefore, God cannot implement a particular action that has a future completion. Once again, to claim that God starts creation at one point and finishes it at another places unacceptable temporal categories on God. Rather, God acts "once", and this one act does not have any temporal duration, nor does it have a teleological intention. This is what it means for God to create eternally. God's activity has no temporal succession, and so does not continue along with the rest of the universe.

―

There is another important dimension of the rejection of temporal language that makes an important contribution to how divine activity should be understood. If God is not temporal, so that God is neither before nor after anything, and neither is divine activity teleological, then God cannot *react* to the universe. Stephen Holmes recognizes this fact. He argues that the divine attribute of "impassibility" "recalls the older sense of 'suffer', meaning to be acted upon". In this way, "God is always active, never passive, God always does, and is never done to" (Holmes, 2011, p. 40). Therefore, Holmes continues that:

> If God is simple and *a se* then nothing outside of God, which is to say nothing in creation, affects God's life in any way; this includes human (or angelic) sin. God is not damaged, lessened or hurt at all by our failures, nor is God restored, repaired or set right in his own gracious act of salvation. (Holmes, 2011, p. 43)

Just as it must be the case that God cannot affect creation, because God is not material, so God cannot be affected by creation, because God is not temporal.

This obviously has massive implications for divine activity. So far this chapter has assumed that divine activity is related to creating, i.e. that God does not create (at least not in an efficient, material sense), because God cannot influence the universe, including provoking the very first moment of space/time (i.e. the Big Bang), because that too must be a material cause.[19] However, the story of divine interaction with the universe arguably focuses more heavily on another divine activity (especially from a Christological perspective), namely, salvation. God cannot save the universe from sin, because that demands that God has a temporal relationship with the universe: in this case that God becomes angry with creation and is then appeased by what Christ (a creature) does.

This also implies the more general idea that God does nothing new. There are many theologians who claim that God deals with the universe in "new" and "different" ways. For Polkinghorne, for example, God would have related to the universe differently just after the Big Bang from the way that God relates to humanity now (Polkinghorne, 2001, pp. 103–5); this has already been rejected in the previous chapter by arguing that there is nothing superior about humanity and that God treats all creatures equally.

Ian Barbour, too, claims that the resurrection is an example of God acting differently, or "doing a new thing" in the universe (Barbour, 2001, p. 20; see also Finlan, 2007, p. 71). The question of the resurrection will be dealt with in a later chapter; however, here it is important to note that the resurrection cannot possibly be an instance of God acting or relating to the world differently, because this would have to imply that God is temporal—acting in one way now and another later. To claim so makes divine activity subject to succession. Even more importantly, the rejection of God doing anything new also means that the incarnation is not God treating the universe any differently; whatever God does it "always" involves the incarnation.

a. Creation as relationship

The rejection of temporal categories in divine activity means that divine activity cannot be efficient causality, as this, by necessity, has to precede other events in order to cause them; efficient causality can only be secondary causality and would mean that divine activity would be univocal with created activity. To claim that God can cause things efficiently would mean that succession would have to characterize divine activity, as it would have to temporally precede them. If efficient causality is necessarily temporal, formal causality is not. Formal causality means that God defines what creation is and grounds it; in this way, divine activity is more akin to a relationship than an act.

This rejection of temporal categories in divine activity, along with the rejection of spatial categories (i.e. God is not material), means that divine activity is not, strictly speaking, activity: it is a relationship. God neither begins to act nor does God finish acting, and that act does not influence the universe. God does not initiate a process, nor does God continue to influence that process; rather, God grounds and defines creation. Thus, William Stoeger writes:

> Creation is not a temporal event, but a relationship—a relationship of ultimate dependence. Thus "cause" as applied to God should be conceived not as a physical force or an interaction,

as it is in physics, but rather in terms of a relationship. (Stoeger, 2010, p. 181)

Creation is not an "event"; it is not something that God *does*. It is a relationship that God has with God's creation: a relationship that is entirely dependent on God, as the Church recognized was so central to Augustine's criticism of Pelagianism. Divine action, therefore, is not, strictly speaking, action: it is the affirmation that God is the ultimate cause of the universe and is solely responsible for it. Maurice Wiles notes that this dependent relationship between God and creatures was essentially the meaning of Aquinas' doctrine of creation (Wiles, 1986, p. 18). Thus, divine activity is not an efficient bringing into being, but a forming of what it means to be created, "after" which creation is able to create itself. God "simply" defines what it means to be created, i.e. mutable and open to failure, a part of which includes being able to "self create".

The eternity of God: the simplicity of God

The non-spatial and non-temporal nature of God, it has been argued, denies that God can influence the world. The non-spatial aspect of eternity has problems with the idea that God, who is immaterial, can have a causal relationship with matter. Likewise, the non-temporal aspect of God has problems with the idea that divine activity can be "localized" and, instead, argues that divine activity must be thought of as a relationship. However, both of these aspects point to another aspect that can also help to clarify divine activity: divine simplicity.

Divine simplicity is not a new doctrine. J. N. D. Kelly claims that the Church Fathers "never tired of pointing out" that God is "simple and indivisible" (Kelly, 1965, p. 254). Unlike creatures that are composed of parts, which are themselves composed of parts and so on and so forth, God is absolutely simple. Indeed, God is not material and so it would not make sense to claim that God is composed of different parts in the first place.[20] However, the simplicity of God's nature also has important implications for God's activity; if God is simple and indivisible then God's activity is simple and indivisible. That indivisibility has two facets, which will be outlined below, and will help to clarify what has already been claimed.

a. Spatial dimensions

From the perspective of the non-spatial dimensions of divine activity, divine simplicity means that God's action "cannot be divided up into discrete phases" (Farrow, 1999, p. 53). This means that God cannot do "many things". As it was suggested in the previous chapter, the fact that humanity can no longer be considered superior or unique means that God does not treat it differently. Therefore, God does not do one thing for humanity and another thing for non-human creatures; divine activity is simple. Thomas Aquinas affirms this simplicity of activity, saying:

> God's activity can be considered either on the part of the doer or of the done. If on the part of the doer, there is only one activity in God ... but considered on the side of what is done, there are indeed different activities. (Aquinas, 1998, pp. 294–5)

This is an important distinction to make. Creatures may be able to distinguish between separate acts, but this does not mean that God is doing different things; these acts are simply different manifestations due to the spatial and temporal nature of existence. Or, put differently, the creature can see different activities, because the creature is also interpreting what God is doing from its own subjective point of view. Augustine appears to argue for the same idea, with a seemingly more explicit reference to the subjective interpretation of divine activity; in his *Confessions*, he writes:

> O truth, everywhere you preside over all who ask counsel of you. You respond at one and same time to all, even though they are consulting you on different subjects. (Augustine, 1991, p. 201)

The subject of God's ability to respond to prayer will not be treated in this book (although the non-influenceability and "non-respondability" of God should suggest how that question would be answered); however, the implication is that God only does one thing. There is only ever one response from God to each and every prayer: Christ.

—

Importantly, the rejection of spatial language on divine activity does not just mean that God treats all creatures identically; it also means that God only does one thing at all times. This means that creation and deification are not distinct acts. God does not do one thing to create and then a different thing to deify. For God, creation and deification are precisely the same act; it is only the creature that distinguishes them. This means that when Maurice Wiles argues for a single divine activity in his book *God's Action in the World* based on a singularity of intention that unites distinct divine acts, this isn't what is meant here; creation and deification are not two distinct divine actions that are united by their intention to create; they are— from God's perspective at least—literally the same act.

Many theologians have made this observation. Teilhard de Chardin, for example, writes that "there is not one moment when God creates and one moment when the secondary causes develop" (Teilhard de Chardin, 1969, p. 23). While this implies that the secondary causes are the same kind of act as the primary cause (i.e. it assumes the univocality of being between divine and created activity), the same essential point is being made here: for God, there cannot be a distinction between creation and deification. For most who recognize this unity of divine activity, the "content" of that divine activity is conservation. Thus, "creation and conservation is one timeless act" (Lane, 1996, p. 45), and "God always and everywhere does the same job of creating-and-upholding [the universe]" (Gregersen, 2008, p. 184).

The reason that this must be the case is "because God is performing one act outside of time" (Haffner, 1995, p. 91). This means that God does not create, nor does God deify (i.e. finish creating), rather God simply conserves, i.e. keeps the universe in being. The universe, it has already been affirmed, can spontaneously create itself due to the fact that creation is always open to mutability; "all" God does is define that mutability and thus keep it in being.

This position can also be seen in the theology of one of the most influential theologians of all time, Pseudo-Dionysius. William Riordan writes that Pseudo-Dionysius "does not seem to distinguish clearly between the gracious generosity of God in creating and what the Scholastics will call habitual or sanctifying grace" (Riordan, 2008, p. 154). Creation, for God at least, is the same as deification. However, what is important about this quotation is the explicit reference to grace. If creation and deification are the same, then there is only one grace. Robert Scuka makes exactly this point. He writes:

> God's prevenient grace is already a salvific or justifying grace in the sense that it is what enables any individual to live life and to respond to God in a faithful way. Thus, God's grace, whether prevenient or salvific, constitutes the condition of, and has the effect of making possible, the very living of life itself … [therefore] the traditional distinction between the prevenient and the salvific grace of God is called into question, for the notion of an additional and distinct salvific grace is superfluous insofar as it makes no sense to suppose that human sinfulness could somehow render the salvific efficacy of God's (prevenient) grace inoperative. Rather, to be meaningful at all, the notion of divine grace must in principle designate something that is universally present and preveniently efficacious, without even the possibility of this efficaciousness being compromised. Thus, what is termed salvific grace is not something additional to prevenient grace, but instead designates the salvific efficaciousness of the one divine grace that functions universally and preveniently. Properly understood, then, "salvific" and "prevenient" do not designate two distinct kinds of grace. They instead designate two dimensions of one and the same divine grace—a grace that in its universal presence is both preveniently and salvifically efficacious. (Scuka, 1989, p. 83)

The length of the quotation here should be taken as a mark of the clarity of his explanation. There is not a "prevenient" grace and a "sanctifying" grace: there is only one grace, which is differentiated "from the perspective of the subject" (Williams, 1999, p. 86), "by the degree of our willingness to receive it" (Hume, 2002, p. 57).

b. Temporal dimensions
From the perspective of the non-temporal dimensions of divine activity, however, while from a "spatial point of view" this means that there cannot be different "kinds" of divine activity, from a "temporal point of view" it also means that there cannot be different "manifestations" of that one activity. God cannot do many things, but God also cannot do many things, many times—nor many things at the same time, nor one thing many times.

The appeal to a single divine act is normally manifested as deism, which, traditionally, is the belief that God does one thing (i.e. create) at the beginning of time and then retreats from contact with the world, never again to intervene. However, this is unacceptable to modern theism as it places inappropriate temporal categories on God.

God cannot do many things, many times, but this does not mean that the single act that God does is restricted to one moment in time. The whole point of the rejection of temporal language is that (drawing on the equation of time and space by Einstein) divine activity cannot be located at a particular point. Traditional deism must also be rejected on the basis that it limits divine activity to a specific moment in time. This is not to mention the criticism that has already been made regarding the fact that even *starting* a "process" must be rejected on the basis that this one act must still be of the same kind as the rest of the process: pushing the first domino is still the same kind of act as the rest of the dominoes pushing each other; divine activity can no more be located at the beginning of time than it can at any point during time. Traditional theism must also be rejected on the basis that it argues for a God who constantly intervenes and influences the world many times doing many different things.

Rather, as it has already been observed, single divine activity is not strictly speaking an act, but an eternal defining of what creation is. Thus, "creation is concerned with ontological origin not temporal beginning" (Craig, 2001, p. 211). The single divine act is a different kind of act—one that defines what it means to be created, "after" which creation can spontaneously create itself out of nothing; creation is *ex nihilo*, not *post-nihilo* (see Verschuuren, 2016, pp. 102–3). Polkinghorne makes this observation when he writes:

> The mystic's God, who is simply the sustaining ground of all being, is not far from the deist's God, whose action is the single creatory fiat by which the world's process is sustained. (Polkinghorne, 1989, p. 36)

God eternally grounds and sustains the world; this is the content of what it means to say that God creates. The temporal simplicity of divine activity seems to support deism, not because God influences the universe only at the beginning of time and then leaves the universe to itself; God "always" leaves the universe to itself. However, divine activity is still singular in nature; God only does one thing, rather than many things, and God only acts once, rather than many times (or continuously).

c. Simple will

The fact that God only does one thing, once (i.e. there are not different "kinds" of divine act, nor are there different "manifestations" of that act) also means that the divine will is simple. Traditional deism has already been criticized, but it must be made clear that arguing for a single, eternal divine act does not mean that God creates "once" with the intention or knowledge that this will lead to the creation of specific things (whether "strongly" in that God intends each individual person to exist, or "weakly" in that God intends some form of intelligent, self-conscious beings to exist). God creates, but God does not create anything in particular. This is the logical conclusion of the rejection of teleology; there is neither direction in the universe, nor inherent values or potentialities (such as consciousness), and God does not create anything specific (i.e. God is not a designer). The "tangled bank" of complexity and diversity of the universe is possible because of God's creating action, but it is also entirely accidental and contingent.

Conclusion

This chapter has argued three important points about divine activity that are seen to be necessary implications of the neo-Darwinian synthesis. In the first place, the accidental nature of genetic mutation and the active prevention of this mutation (along with the sole sufficiency of it to account for all variation) means that there is not just an absence of evidence of divine interaction with, and influence on, the universe, but there is evidence of absence. However, if the classical, traditional definition of God that has characterized the greatest theological minds (such as Augustine, Pseudo-Dionysius, Anselm, and Aquinas) is considered, then arguing for the distinctness of God from creatures (i.e. God and creatures are not univocal) means that this is what should be expected from a theory of divine activity. God cannot directly influence the world. To claim that God can influence the world is to reduce divine activity to created activity (i.e. secondary causality).

The same arguments made regarding the rejection of spatial categories for God are equally applicable for temporal categories, due to the fact that space and time are inextricably and inherently linked. If God cannot be located spatially (i.e. there is no evidence of divine influence), then God cannot be located temporally. As it was argued in a spatial context, any suggestion that God can influence the world would be to pinpoint divine activity spatially and temporally, but this is not what is to be expected. Rather, divine activity is not an act at all: it is a relationship in which creatures depend entirely on God for their existence.

Both spatial and temporal rejection of divine activity have another important implication: namely, that divine activity must be simple. God does not do many things, otherwise divine activity could be divided. Likewise, God does not do one thing many times; God does one thing once. This one thing, however, is not an act that can be located at any point in space/time (such as creation or deification); this one divine activity is an eternal activity. That eternal activity is the eternal definition of what it means to be a creature (i.e. conservation); such a definition of creatures as being mutable (i.e. that which the neo-Darwinian synthesis observes and describes) means that the universe can create itself without the need for a divine efficient cause (which would make God and creatures univocal).

This completely rejects theism, precisely because theism claims that God *continually* and *directly* (i.e. temporally and spatially) influences the world. Likewise, it rejects traditional deism, because this claims that the single divine act is understood as efficient causality (i.e. at the beginning of time). However, deism is at least correct in its insistence that God only does one thing once. The only problem with traditional deism is that it applies temporal categories to God by claiming that God starts and finishes creating and then does nothing. Yet, this chapter has argued that it is necessary to see divine activity as eternal, without beginning, without end and without succession. Divine activity is a relationship (i.e. formal/primary), not an act (i.e. efficient/secondary).

However, a relationship still needs a connection. If God and creation are completely equivocal, completely differently different and otherly other, then this hardly speaks of relationship. This chapter (and neo-Darwinism), therefore, does not quite argue that God cannot influence the universe: it argues that if God wants to influence the universe God needs to be created. Somewhat contradictorily, God still needs a temporal and spatial locus in order to be in relationship with creatures and to define what it means to be a creature.

Fortunately, Christian theology already has a doctrine that can deal with this precise problem, namely, how it is that God—completely and utterly transcendent and incapable of influencing and interacting with creatures that completely depend on God for their existence—can still be the ultimate ground of creation: Jesus Christ.

CHAPTER 4

The Person of Christ

In the previous chapter, I argued that theologians should not expect God to influence the world, and, audacious as this claim sounds, it is nothing but a translation of traditional classical theology of the transcendence of God into modern, neo-Darwinian language. However, if God is solely and uniquely responsible for the universe, and the universe exists, then there must be divine influence. Nuancing that conclusion, it means that the non-spatial, non-temporal, simple, and primary divine activity that is the grounding, conserving, and defining of what it means to be created must have a temporal and spatial locus.

Christianity already has a solution to this problem in the person of Jesus Christ, who is both fully God and fully human. Divine activity, in other words, is a uniquely Christological concern: the second person of the Trinity is the unique agent of divine activity.

In the same way that the conclusion of the previous chapter—that God cannot influence the universe—is nothing but a translation of traditional theological categories, so this notion that Christ is the solution to the problem of the complete transcendence of God is not a novel solution. Frances Young writes:

> [God] was impassive and unaffected by anything external. He could have no history or development, no involvement. The consequences of such a concept was [sic] that it was hard to relate God, or the One, with the multiplicity of things, the world of which he was supposed to be the source and ground of being. His utter transcendence meant his substantial irrelevance to the problem of which he had originally been the solution ... inevitably the solutions involved some kind of system of mediators

> of a "hierarchy of being" linking the ultimate transcendent one, who was even "beyond being", with the known world (Young, 1977, pp. 24–5)

The same problem that neo-Darwinism and modern science creates for the belief in God is precisely the same problem that Greek philosophy created for early Christian belief: how can God, who is completely transcendent and impassable, relate with that which is passable? This adds further evidence to the claim that was argued in the previous chapter about the transcendence of God and God's lack of involvement or "substantial irrelevance". It is here, then, that Christ became the solution. Young continues that:

> Logos theology and Trinitarian doctrine made it possible for God to be *involved*. The impassable, transcendent *one*, beyond being, was intellectually adequate and mystically inspiring, but could not elicit the faith and devotion of most ordinary mortals. The doctrines of the Logos and the Spirit made it possible to believe in a God who is both transcendent and immanent, however paradoxical that might seem to be. (Young, 1977, pp. 41–2)

This introduces the theme of this chapter: that Christ is the sole agent of divine activity—not because Christ is the spatial and temporal point of God who pushes the first domino, but because Christ is the definition of what it means to be created (i.e. Christ is the relationship of dependence argued in the previous chapter). God does not need to be created so that he can initiate the "process" of the universe (i.e. efficient causality), but God needs to be created in order to define what it means to be created (i.e. formal causality), which is conservation. Christ is the "firstborn of creation" and "in him" all things were created (Colossians 1:15–16). What Jesus Christ accomplishes in and through his incarnation is not just "the supreme example of God's action in the world" (Wiles, 1986, p. 82), it is the *only* example of God's action in the world.

The hand and word of God

The idea that Christ is the agent of divine activity is not a novel claim; it has a basis in both biblical and Patristic theology. According to Ellverson, Gregory of Nazianzus, for example, acknowledges that God the Father is "the initiating one", but Christ is "the one who does the actual creating" (Ellverson, 1981, p. 90). For many Patristic authors, the claim that Christ is an agent of divine activity takes the shape of giving Christ the title of "hand of God". Irenaeus, for example, writes:

> For by the hands of the Father, that is, by the Son and the Holy Spirit, man, and not [merely] *part* of man, was made in the likeness of God. (Hall, Rae and Holmes, 2010, p. 284)

Irenaeus is clear here: all divine activity comes from the Father, but it is mediated to the universe through the Son and the Spirit.[21] Even though Robert Barron notes that the intention behind Irenaeus' claim is "to designate consubstantiality" rather than "mere instrumentality" (Barron, 2015, p. 51), that instrumentality is certainly an important implication of consubstantiality. It is precisely because Christ is consubstantial with God that he can be an agent of divine activity. In fact, this very point forms the basis of Athanasius' most famous and influential work, *De Incarnatione*, in which he argues for the divinity of Christ from the fact that Christ performs a work that is the prerogative of the divine, namely, re-creation (Athanasius, 1954, p. 56). The *very* reason for arguing for Christ's divinity in the first place is to make sense of his being the agent of divine activity; Christ cannot do divine things if he is not divine himself.

Irenaeus is not the only Patristic writer to understand Christ as the "hand of God". Augustine also uses this image to claim that Christ is an agent of divine activity. O'Connell notes that Augustine understands "Scripture's evocations of God's hand or arm" as "referring to his creative word" (O'Connell, 1994, p. 110). What is important about Augustine's identification of God's hand with Christ is that he applies this ascription to Old Testament passages. For Augustine, therefore, whereas the incarnation might represent God relating to the world in a new way, there is still a sense in which Christ is also the agent of divine influence *before* the incarnation. Christ is not just responsible for *some* divine activity: Christ is responsible for *all* divine activity. Christ's response to the Jews that "I am" (John 8:58) also makes this connection. Whether or not Christ is doing something new in the incarnation (I argue that he is not, as this places unacceptable temporal categories on divine activity), Christ is still responsible for divine activity before the incarnation. However, nowhere is the connection between the incarnation and divine activity "before" the incarnation greater than in the image of Christ as "word of God".

—

Whereas the Fathers used the image of the "hand of God" to argue that Christ is an agent of divine activity, the Bible uses the image of the "word" with, arguably, much greater force. The opening chapter of the Gospel of John is the *locus classicus* of this biblical theme in which the author writes:

> In the beginning was the Word, and the Word was with God and the Word was God. He was in the beginning with God. All things came into being through him, and without him not one thing came into being. What has come into being in him was life, and the life was the light of all people. (John 1:1–4)

For the author of the Gospel of John, it is clear that God creates through speaking, or, as Bianchi puts it, "when God speaks, things come to be" (Bianchi, 1998, p. 24). Divine speaking is the biblical image of divine action. Strengthening this claim, as many commentators have noticed, is the fact that John's opening chapter quoted here is an "explicit connection" (Maloney, 1968, p. 75) to the opening chapter of Genesis, in which God speaks the world into being (although Pelikan argues that John 1 is actually a "Christian gloss upon chapter 8 of Proverbs", which itself is a gloss on Genesis [Pelikan, 1971, p. 139]).

The implications of such an allusion are astonishing, and theologians rarely give full appreciation to this remarkable assertion. The author of John is claiming that the human Jesus is none other than the word of God that is responsible for the creation of the universe. "God's word is

God's act", writes Wiles, and this word is explicitly identified with the person of Jesus Christ.

Other New Testament texts (some probably written before the relatively late fourth Gospel) also see in Christ the agent of divine activity and link the person of Jesus Christ with the initial act of creation. Paul's use of the title "Lord" (e.g. 1 Corinthians 8:4–6), for example, as a "substitution for the divine name of the one God", shows that he understood Christ as being identified with the creating God and thus "present at creation as its mediating agent" (Wright, 2002, p. 68). However, there is little doubt that the Epistle to the Colossians is Paul's most important passage in this regard:

> [Christ] is the firstborn of all creation; for in him all things in heaven and on earth were created, things visible and invisible ... all things have been created through him and for him. (Colossians 1·15ff.)

As with John, although without the explicit reference to the opening chapters of Genesis, the person of Jesus Christ, who became incarnate and "dwelt among us", is the agent of divine creation that is responsible for the creation and conservation of the whole universe. It is noteworthy here that there is a connection with primary cause that was outlined in the previous chapter; Christ is not the "firstborn of all creation" in a temporal sense but in an ontological one. Christ does not precede all creatures temporally (i.e. efficient causality), but ontologically (i.e. formal causality); it is from Christ that all creatures obtain their being. Christ is creator in a primary, formal sense—i.e. *ex nihilo*—not in a secondary, efficient sense—i.e. *post nihilo*.

It is not just John and Colossians that add support to the claim that Christ is the "fontal source of all grace" (McCord Adams, 1999, p. 20), i.e. the sole agent of divine influence on the universe. O'Collins notes that Hebrews 1:3 and 1 Corinthians 8:6 also "confess [Christ] as the universal and exclusive agent of creation" (O'Collins, 1995, p. 302). It is especially noteworthy that O'Collins calls Christ the *exclusive* agent of divine activity. There is only one word, Christ, which God eternally speaks. However, Christ's claim that "no one comes to the Father except through me" (John 14:6) and the author of the Letter to Timothy's similar claim that "there is one God; there is also one mediator between God and humankind, Christ Jesus, himself human" (1 Timothy 2:5), further points to Christ as the unique agent of creation.

Moreover, there is here an implication that supports the conclusions for which I argue. As Shortt explicitly notes in relation to the idea of Christ as mediator, "the gap between heaven and earth is only bridged definitively in the figure of Christ" (Shortt, 2016, p. 77); therefore, if "no one comes to the Father except through [Christ]' (John 14:6), then this implies that the Father also comes to no one except through Christ. After all, Jesus exclaims in the Gospel of John, "I have come down from heaven, not to do my own will, but the will of him who sent me" (John 6:38). God cannot influence the universe outside of Christ (and that influence is simply the conservation of all creatures by grounding and defining what creatures are). This same non-influenceability of God and Christ as the sole agent and vehicle of divine influence can also be inferred from Thomas Merton, who writes of the role of Christ:

> God is everywhere. His truth and his love pervade all things as the light and the heat of the sun pervade our atmosphere. But just as the rays of the sun do not set fire to anything by themselves, so God does not touch our souls with the fire of supernatural knowledge and experience without Christ. (Merton, 1961, p. 106)

God cannot influence the universe; the divine influence must be mediated through Christ.

Although the image of Christ as the "hand of God" was not something that the Patristic writers found in the Bible, they did pick up on the image of Christ as the "word of God" (which achieved the same purpose as the image of the "hand", i.e. Christ as agent of divine activity). Irenaeus, for example, writes that "there is one God, who by his Word and Wisdom made and ordered all things", explicitly identifying this "Word" with "our Lord Jesus Christ" (Irenaeus, 1969, p. 76). Clement of Alexandria, too, in *Protrepticus*, writes:

> The word who in the beginning gave us life when he fashioned us, as creator, has taught us the good life, as our teacher, that he may afterwards, provide us with eternal life (Clement of Alexandria, 1969, p. 172).

For both Irenaeus and Clement (as examples of the wealth of Patristic literature), the person of Jesus Christ is the creator and sustainer of things. Interestingly, Tertullian not only supports this idea, but expounds it in ways that come remarkably close to how this chapter understands Christ as divine agent. In *Against Praxeas*, he writes:

> Is it conceivable that an insubstantial being should have created concrete things; a void, solids; an incorporeal, bodies? For though it may sometimes happen that something can be made whose nature is different from that of the agent of its creation, it is still true that nothing can be created by the agency of what is insubstantial and void. How then can the Word of God be insubstantial and void, that Word who is called the Son of God, and given the name of God? (Tertullian, 1969, p. 104)

It would be a mistake to claim that Tertullian, who lived some 1,500 years before the invention of modern science, is here arguing for precisely the same thing for which I argue, but there are remarkable similarities. Translating what Tertullian expounds into the language of modern science would yield nothing but the conclusion of the previous chapter; if the transcendent God (who is equivocal with creation) wants to influence creatures (including creating them in the "first" place), then God must become created (i.e. univocal with creation), which God achieves through Christ. "God could not create unless God was incarnate" (Delio, 2013, p. 127).

Communicatio idiomatum

If God wants to influence the universe, then God must become created. However, this leads to the question of the relationship between the divine and the created *in* Christ: how is it that the created nature of Christ can perform divine activity? The answer to that question is the doctrine of the communication of properties, or *communicatio idiomatum*.

The origin of this doctrine lies in the fifth-century debate over the application of the title *theotokos* to Mary, which led to the third ecumenical council at Ephesus in 431. Nestorius disagreed with applying the title to Mary, because he claimed that it made God a creature and denied the transcendence of God. However, Cyril of Alexandria criticized Nestorius, as the rejection of the title *theotokos* seemed to separate the divine from the humanity in the person of Christ. The two natures in Christ may well be distinct (they do not blend into a "third nature" as Chalcedon would uphold in 451), but this does not mean that they are not fully united in the one person of Christ.

As O'Collins puts it, "despite the duality of natures, there is only one subject of attribution" (O'Collins, 1983, p. 183). Thus, if the first council (held at Nicaea in 325) upheld the full divinity of Christ, and the second council (held at Constantinople in 381) upheld the full humanity of Christ, then the third council upheld the full unity of those disparate natures in the one person of Christ.

This full union of natures in the one subject or person of Christ means that what can be said of one nature can be applied to the other. Or, more precisely, as Cross notes, the communication of properties is not the "ascription of the properties of one nature to the other", which would entail a change in the divine nature, but is "the ascription of divine and human properties to the person of the word" (Cross, 2002, p. 184). In other words, the human nature does not become divine, and the divine nature does not become created, but the one person of Jesus is both. This means it can be claimed without contradiction that God died on the cross and the human Jesus created the universe. This does not mean that the divine nature can die, but that the one single subject, who was both divine and created, died and, because of the union of natures in Christ, this can be said of the divine nature. Thus, Thomas Merton can claim, quite correctly, "that the human Christ is God" (Merton, 1961, p. 107).

It may very well be the case, as the theory of "reduplication" claims, that the divine properties applied to the person of Christ are applied to Christ *qua* God and, likewise, the human properties that are applied to Christ are applied to Christ *qua* human, but this does not mean that these statements do not refer to the same *identical* subject, who acts as both.

This should not be taken as a criticism of the "reduplication" attempt to solve the problem of the relationship of disparate properties in Christ. It simply means that "reduplication" is only half of the story. It is obvious that the union of the natures in the person of Christ does not change those natures—such that the divine nature becomes mutable—but it means that those disparate natures are fully united in the one person of Christ. "The human nature does things," writes Cross, "but the actions of the nature are predicated of the divine person" because of *communicatio idiomatum* (Cross, 2002, p. 220). John of Damascus supports this idea, saying:

> Thus, we do not say that the operations are separated and that the natures act separately, but we say that they act conjointly, with each nature doing in communion with the other that which is proper to itself. (John of Damascus, 2013, p. 158)

The divine and created natures in Christ are still distinct, and still have their own will and energy that is distinct from the other, but both natures act in complete communion: "the action of God in Christ comes through the acts of the human Jesus" (Wiles, 1986, p. 87). As Aaron Riches writes:

> The hypostatic singularity of Jesus, who is Logos, means that the Chalcedonian language of "without separation" must apply to the energies of Christ as well [as the natures]. The divine energy must be synergistically "one" with the natural energy of this human being. (Riches, 2016, p. 145)

Just as there is a union of natures, so there is a union of energies. Christ does not act as God at one moment and act as a creature at another; Christ always acts as God-creature.

Again, it is important to emphasize, this does not mean that the human Christ's divine activity is efficient or secondary, but that in the person of Christ, precisely by being united to the creaturely nature, the divine nature acts and influences creation by sustaining it in being; the divine act that the human Jesus does is define what creation is (i.e. mutable and open to failure), by being the firstborn of creation, and "once" God defines creation as such, the universe can spontaneously create itself from nothing (and because the divine act is eternal [i.e. timeless] it doesn't have to "precede" the beginning of the world).[22] As Riches puts it, the incarnation is "the 'site' of eternal happenings" (Riches, 2016, p. 43).

—

Many theologians claim that the communication of properties is an incoherent doctrine because it means ascribing opposite characters to Christ, such as omnipotence and limited power, or omniscience and limited knowledge (O'Collins, 1995, p. 234). Calling it the "incoherence problem", critics of the incarnation (at least as it is traditionally understood as outlined here) claim that it is inherently illogical, as it means that Christ would have to have opposite capacities at the same time.

However, this can be solved in two ways. Firstly, the equivocality of divinity and creatures—more specifically that God is nothing and has "no properties"—means that there is no contradiction, because there cannot be any category that "precedes" God and can be applied to both God and creatures (i.e. a comparison can only be made between two things that have "some common respect" [Turner, 1998, p. 42], which is more often than not "being").

As has already been noted, God cannot be omniscient, because that would mean that the category of knowledge could be applied to both divinity and creatures, which would make them univocal. If divinity and creatures are equivocal (as the previous chapter suggests is the implication of neo-Darwinism) then divinity and creatures cannot be in "competition". Therefore, it makes no sense to claim that, for example, the divine nature's immutability is in competition with the created nature's mutability. Thus, Richard Norris writes:

> The *difference* between God and humanity is a matter neither of contrariety nor of contradiction, that God is not related to us as an element or factor or reality that is either interchangeable with the creature as contrary (i.e. a different thing of the same general sort) or incompatible with the creature as its utter negation ... [thus] maybe after all, suggests Chalcedon, God and humanity are not related on "yes" and "no" or "off" and "on". (Norris, 1996, pp. 157–8)

In this way, the incoherence problem is solved. There is no incoherence, because the divine and created properties only lack coherence if there is a category to which both creature and God belong. Or, to put this more simply, the "incoherence problem" is only a problem if God and creatures are univocal.

Secondly, if the doctrine is understood as a doctrine of activity rather than a doctrine of being, then the problem can be seen in a different light. Thus, Riches writes:

> The question of how the one who is "beyond being" can become a "human being" moves, in this way, from a question of being as such to the question of the eventual dynamic of action [i.e. as an active divine act or event]. (Riches, 2016, p. 143)

The communication of properties is not so much a doctrine of *who* Christ is (although this is, of course, an important consequence) as it is a doctrine of *what* Christ does. When *communicatio idiomatum* is understood in this way, the important consideration is not about how the divine and created natures "co-exist" in the person of Christ, but how the person of Christ is responsible for both divine and created activity. The *communicatio idiomatum*, then, claims that the actions of the human Christ can be an expression of the divine energy as much as of the human energy. To put this slightly differently, the question is not, for example, how the person of Christ can be both mutable and immutable, but how the mutable action of the created body of Christ can manifest the immutable divine energy.

This does not mean that there is not a communication of properties in the person of Christ; there must be a union of being for Christ to be a divine actor, as Athanasius made clear, i.e. it is not a description of how the human Jesus carries out divine activity without that human Jesus also being fully divine. However, if God and creatures are equivocal, so that there cannot be anything that can be descriptive of both God and creatures (such as knowledge), then *communicatio idiomatum* is better understood as how the one person of Christ achieves the divine act of conservation through the human Jesus.

This emphasis on *communicatio idiomatum* as a doctrine of action rather than ontology could be seen as an interpretation of Aquinas' claim that the human nature is a "divine instrument". Riches writes:

> The [Thomist] language of *instrumentum divinitatis* carefully signifies the absolute *unio* of operation of divinity and humanity in the incarnate son by stipulating that the integral human nature only "acts" insofar as it acts as "one" with the divine logos ... in Christ, whatever distinction we must make in order not to confuse the natures, in all cases it is "one and the same subject", the one Lord Jesus Christ, who acts and wills in the perfect unity of his being... [thus] "[Christ's] divine operation uses his human operation, and his human operation participates in the power of the divine operation". (Riches, 2016, pp. 182–3)

The person of Jesus Christ, who is the perfect union (without confusion) of full divinity and full humanity, performs the eternal divine activity. God *uses* the human nature of Jesus to influence the world (i.e. the simple, non-temporal act of conserving the universe in being).

Claiming that Christ's human nature is an instrument of the divine does not mean that Christ's human nature is a puppet; this is not a return to Apollinarianism (in any case, this book has denied the existence of a soul and has denied that God can have a mind, which could "replace" the human soul or mind in Christ). More specifically, the claim that the created nature of Christ is an instrument of the divine does not deny that

Christ has a created will or energy. What it means more precisely is that the created nature, because of *communicatio idiomatum*, is the agent of divine activity, which is the conserving and grounding (both of which are manifested as a defining of creation) of the universe (i.e. it is formal activity, not efficient activity).

—

It is important to note here that the use of *communicatio idiomatum* as a theory of divine activity assumes that there is a comparable union of the persons of the Godhead to the union of natures in the second person. The human person of Christ is *only* the second person of the Trinity, but he is the agent of the activity of *all* three persons. While very few, if any, theologians see Christ as the *sole* and *unique* agent of divine activity, the unity of the Godhead that many theologians do note could be seen as support for this view. Lossky, for example, argues that the possession of one nature by the Godhead means that the Godhead have one "single will", one "single power" and a "single operation" (Lossky, 1957, p. 53; see also Pelikan, 1977, p. 78). Nicholas Cabasilas, also an Eastern theologian, likewise writes that the Godhead "performs all things by one power" (Cabasilas, 1974, p. 74).

However, more pertinent to the direction this book has taken, J. N. D. Kelly, by noting that the three persons of the Godhead are "possessed of one and the same activity", explicitly links this to the fact that there is "a single grace which is fulfilled from the Father through the Son in the Holy Spirit" (Kelly, 1965, p. 258). This means that while I will argue that *only* Christ creates (because it is only Christ who becomes incarnate [Meyendorff, 1964, p. 231]), the whole Trinity are necessarily included in that creation, because there is only "one" divine activity. Arguing that only Christ acts does not mean that the Father and the Spirit are not included in that action: Christ does the work of the Trinity.

The analogy of being

If the Patristic *communicatio idiomatum* must be nuanced to show that it is actually a description of divine activity in Christ, because the two disparate natures in Christ are equivocal, then the Scholastic analogy of being can provide the ontological framework on which the theory of divine activity can hang.

The claim that Christ is the "one mediator between God and humankind" (1 Timothy 2:5), such that in Christ there is a meeting of divinity and humanity that allows the equivocal divinity to influence the universe, provides a nuanced reasoning for the analogy of being. So far, the solution to the Thomistic rejection of univocality *and* equivocality of being has been left open. It has been explained sufficiently why the first option, univocality of being (espoused by the medieval Franciscans), must be rejected, as this places God within the causal nexus of the universe and thus demands that God is material (i.e. created rather than creator). However, for Aquinas, the second option, equivocality of being, had to be rejected as well because this would mean that God could not be responsible for creation. If humanity was created in the image of God, then, reasons Aquinas, humanity and God could not be equivocal either.

The previous chapter, drawing on the evidence against divine influence provided by neo-Darwinism to accentuate and nuance the transcendence of God, argued that God and creatures must be equivocal. The appeal to analogy—as opposed to univocality or equivocality—was made as a "middle way" (Ogden, 1983, p. 86), because neither could adequately describe the relationship between God and creatures. However, now that creation is understood to happen solely and uniquely through the relationship of divine and creature in the one person of Christ (i.e. the communication of properties), the analogy of being can be nuanced.

Aquinas is surely correct that the analogy of being must describe the relationship between God and creatures. But, now, this is only because of the union of those natures in the one person of Christ, without which there could be no meeting of God and creatures; this is because without the meeting in the person of Christ, God and creatures would be equivocal. The analogy of being is not a philosophical problem, as it was for Aquinas and the Scholastics—it is a Christological problem. As

Bernard Montagnes points out, the analogy of being applies to distinct natures that "are all in a relation to the one among them to which the common meaning primarily belongs" (Montagnes, 2004, p. 26), but that relation only happens in Christ. The relationship between God and creatures can only happen in the person of the second person of the Trinity, whose full unity of natures makes Christ the "one mediator".

In other words, the analogy of being does not describe a general relationship between God and creatures; it describes the relationship between divine and created natures in the person of Christ. God and creatures are equivocal, but because of the full union of those equivocal natures in the one person of Christ, there can be a meeting between them, and therefore a real connection, which leads to analogy. O'Collins seems to hint at this Christological solution to the analogy "problem", when he writes:

> Without the personal self-manifestation of God that the incarnation brings, God could seem remote and distantly other, even separated from us by an infinite gulf. (O'Collins, 2002, p. 26)

To put that same point in the language of analogy, without the personal union of natures in God, then God can only be equivocal and transcendent, unable to meet with and influence creatures; there is only an analogy of being because the otherwise equivocal divine and created natures are fully united in the one person of Christ. Robert Barron, too, hints at such a solution to the question of the Christological dimension of the analogy of being in his book *The Priority of Christ*. The "priority" of Christ, Barron argues, is an "epistemic priority" in which the universe is only understandable in the light of Christ; outside of Christ there is no "natural theology" (i.e. that knowledge of God can be gained through reflection on God's creation, without the revelation of God in Christ [see Torrance, 1976, p. 1]).[23] He writes:

> Just as the Logos of the Father is the power through which all things are intelligible, so that same Word is the power through whom all things are known. Accordingly, it is in Christ and through Christ that even the simplest act of cognition takes place;

"natural" reason is thoroughly Christological . . . what we know and how we know is conditioned by what was revealed in Jesus Christ. (Barron, 2007, p. 148)

While Barron does not comment on the wider implications of his idea, I argue that there is an ontological complement to it. The world is only intelligible in the light of Christ because the world was created only through and in Christ; it is not just reason that is "thoroughly Christological", the universe itself is "thoroughly Christological".[24] In much the same way that Aquinas denies the equivocality of being, because creatures are created in the image of God, so Barron argues that the universe is only intelligible in the light of Christ, because it is Christ who is responsible for its creation.

The analogy of being, then, as Aquinas correctly argues, is a doctrine that describes the ontological relationship between God and creatures. However, his exposition of this relationship must be criticized. It has been outlined that, following from the biblical and Patristic claim that Christ is the agent of divine activity, the neo-Darwinian synthesis suggests that Christ must be the *sole* agent of divine activity. This means that the relationship between God and creatures that is responsible for keeping them in being (i.e. divine activity) cannot happen outside of Christ. The analogy of being, therefore, can only be a description of the relationship between God and creatures *within* the person of Christ. Outside of Christ there is an "infinite distance [or 'infinite gulf'] between God and creation" (Florovsky, 1976, p. 46)—i.e. God and creatures are equivocal.

The incarnation

However, despite the arguments proposed in this chapter regarding *communicatio idiomatum* and the claim that Christ is the sole and unique agent of divine activity, this does not mean that such activity is *efficient* causality. Divine activity is still primary, rather than secondary; it is still formal, rather than efficient; and it is still relationship, rather than act. All this chapter has claimed is that the person of Jesus Christ, who is both fully divine and fully created, is the locus of this relationship. The singular person of Christ, in which divinity and creation are united, overcomes the problem of equivocality of divinity and creation, and solves the problem of how the divine (which is utterly transcendent of the universe and thus incapable of influencing it) can influence the universe.

This means that Christ does not become created in order to act as God: the becoming created *is* the act of God. God does not become created and then conserve; the becoming created is the act of conservation. The incarnation, then, is not the means to an end; the incarnation is an end in itself. The incarnation is not the way that God can become created in order for divine activity to become secondary and thus able to influence the universe; the incarnation is the way that God initiates the relationship with creatures. It is to this idea that I now turn.

CHAPTER 5

The Incarnation

In the previous chapter it was shown that the person of Jesus Christ solves the problem of how the transcendent and eternal (i.e. non-spatial, non-temporal, and simple) God can influence the universe. The hypostatic union of the divine and created in Christ is the sole and unique meeting place of God and creation. The relationship of dependence that keeps creation in being is found in the relationship between the divine and created natures in the person of Christ. This union overcomes the equivocality of being that is necessitated by the absolute transcendence of God that the evidence of neo-Darwinism advocates.

By drawing on a nuanced interpretation of the Scholastic theory of analogy of being, the two natures in Christ, whilst still distinct, are united in the single subject of Jesus Christ. This also draws on the Patristic theory of *communicatio idiomatum*, further accentuating that the two distinct natures in Christ are in full union, and emphasizing that, while they are distinct, the divine eternal activity (i.e. conservation/grounding of being) is undertaken by the human Jesus.

This emphasis on the hypostatic union as a way of solving the problems that neo-Darwinism creates for divine activity obviously puts a further emphasis on the incarnation. The incarnation, as an event (i.e. the literal act of assuming a created nature, not incarnation as a "general" term to describe the whole of Christ's created life), is the way that this hypostatic union comes about. This chapter will discuss how the incarnation contributes to and furthers this interpretation of divine activity through the person of Christ.

The incarnation as creation

While it may be somewhat simplistic, it is not altogether impossible to demarcate between two distinct Christologies. The Western Christology, influenced by the Augustinian theory of original sin, understands the role of Christ in almost exclusively salvific terms: Christ saves. On the other hand, the Eastern Christology, less constrained by a strict interpretation of original sin, is also less constrained in understanding the role of Christ. For the Eastern Christology, Christ's role of saviour is seen as part of the wider creating action of God. In other words, the preoccupation of the West with the problem of sin constricts its Christology.

For the Western Church, the incarnation (as an event) was simply the means that allowed the crucifixion to happen. However, for the Eastern Church, the incarnation takes on a much more central role, with the crucifixion seen within the light of the incarnation. As many theologians have noted, it is not *what* Christ does but *who* Christ is (Williams, 2007, p. 38). So much so, in fact, that Michael Winter claims that many of the Eastern Fathers "speak of the incarnation itself as the sufficient cause of the redemption", rather than confining redemption to the crucifixion (Winter, 1995, p. 40).

―

So far, this book has presented a number of reasons why the Western interpretation of salvation should be criticized and the Eastern emphasized. The neo-Darwinian synthesis rejects the historical accuracy of the Genesis narrative, upon which the Augustinian theory of original sin is pinned. Lyell's geology, which claimed a much older world than the Genesis narrative, provides strong evidence for the rejection of the Genesis narrative. Darwin's theory of evolution through natural selection also rejects the Genesis narrative, disagreeing with it on *how* humanity came into being. Furthermore, not only did the rediscovery of genetics in the twentieth century solidify and strengthen Darwin's disagreement, it also contributed to that disagreement by providing evidence against monogenism and the idea that the whole human race descended from one couple.

The rejection of the Genesis narrative has the knock-on effect of removing the one historical event upon which the Augustinian theory of original sin is pinned. As it has already been argued, this all means that the category of salvation as a way of interpreting the role of Christ must also be rejected. Christ does not save humanity, because there is nothing from which humanity (or any other creature) requires salvation.

However, neo-Darwinism is not the only reason why the paradigm of salvation should be criticized. As Chapter 3, "Divine Activity", made clear, it does not make theological sense to claim that God can save, if that claim to salvation requires God to react to historical events.[25] If God is the creator of time (by being ultimately responsible for it, even if not immediately and directly responsible for it), then God must be outside of time and therefore cannot be susceptible to succession, which is characteristic of temporality. If God is not susceptible to succession, then divine activity cannot be reactive, which, by its very definition, has to happen after other events. The eternity of God, therefore, from a purely theological perspective, rejects the category of salvation, because it can only be applied to God along with temporal categories. Christ does not save humanity, therefore, not only because there is nothing from which humanity requires salvation, but also because God cannot "save" in the first place.

Thus, neo-Darwinism, by contributing to both anthropology and theology, provides a number of different reasons why the category of salvation must be rejected. Drawing on the Eastern interpretation, the role of Christ is not limited to the narrow role of saviour, but is widened and broadened to be more inclusive. This Eastern interpretation, according to Georges Florovsky, finds that an "adequate" answer to the reason for the incarnation "can be given only in the context of the general doctrine of creation" (Florovsky, 1976, p. 170). The role of Christ is to be found in the divine act of creation.

However, Eastern theologians, while they do see the incarnation within the context of creation, do not equate the "initial" act of creation with the incarnation. Rather, the incarnation represents a continuation of the process that was started with the "initial" act of creation and the means by which it is completed. The incarnation is not how creation comes into being: it is God doing something new in creation to bring it to greater fulfilment. Thus, Andrew Louth writes:

> [For Eastern theology, there is one] arch stretching from creation to deification... [and] a lesser arch leading from fall to redemption, the purpose of which [is] to restore the function of the greater arch. (Louth, 2007, p. 34)

The Fall and redemption are only a minor setback that temporarily distracts from the completion of creation, which is deification. Deification, Louth continues, "exposes the full extent of the consequences of the incarnation" (Louth, 2007, p. 34). While for Louth the incarnation may still have some salvific overtones, what is important is that this is not a return to an original perfection; rather, the incarnation contributes to the continuing completion of creation.

Such an idea has already been noted as being found in the early Fathers, such as Irenaeus, who expressed creation as a continual, ongoing phenomenon, and not something that was completed and subsequently tarnished. Thus, Irenaeus "always sets Christ's coming against the backdrop of creation" (Wilken, 2003, p. 66). Christ comes, not to save and restore, but to complete and deify. Deification becomes the end of creation, and it is this that Christ ushers in through his incarnation. The incarnation is not salvific: it is creative.

Many other theologians express this point. Mantzaridis, for example, notes that the result of the hypostatic union is the "deification of the human nature he assumes" (Mantzaridis, 1984, p. 29). For him, importantly, there is a link between the deification and the assumption of human nature. This will be explored below, but it is important to note here that it is not only that the incarnation as a general category (i.e. the whole human life and ministry of Jesus) deifies humanity, but also it is the incarnation as a specific event (i.e. the assumption of human nature) that deifies. That is, the incarnation is not a means to another end, but is an end in itself.

In fact, it is not, strictly speaking, exclusively limited to Eastern theologians to expound this interpretation of the role of Christ. Some westerners were also unhappy with the narrow constraints of a legalistic

Western theology and sought to widen the scope of who Christ was and what he achieved.

Duns Scotus, for example, famously disagreed with Augustine's negative answer to the now classic question of whether Christ would have become incarnate had Adam not sinned. Florovsky notes that for Duns Scotus, the idea of Christ becoming incarnate without Adam's sin was not a mere theological frivolity or theological luxury, "but rather an indispensable doctrinal presupposition", the main emphasis of which is to see the incarnation "in the total perspective of creation" (Florovsky, 1976, pp. 165–6). Francis Klauder makes the same point by claiming that Aquinas could consider creation without the incarnation, but not the incarnation without redemption, whereas Scotus could not conceive of creation without the incarnation, but could conceive of the incarnation without redemption (Klauder, 1971, p. 80). In the twentieth century, Teilhard de Chardin combined this view with the theory of evolution.

The role of Christ, then, is deification. However, the Eastern theologians did not just shift the reason for the incarnation from redemption to deification (or perhaps *resist* the shift that the Western theologians made from deification to redemption), they also shifted the focus from the crucifixion to the incarnation. For the Eastern theologians, it was the incarnation itself that was efficacious, and this reveals something important about how they understood divine activity.

a. The divine "mechanism"
As it has already been suggested, deification is not to be thought of as the overall consequence of Christ's life, but is rather the result of the incarnation *as an event*. Regardless of what Christ does with his life, he has already achieved deification simply by assuming a created nature in his person; it is not *what* Christ does, but *who* Christ is (i.e. a union of divine and created natures). For the Eastern theologians it is the assumption of created nature by the divine person of Christ that brings about deification, not anything that Christ does subsequently.

In this way, Gregory Palamas, an immensely important Byzantine theologian, asks whether the deification and transfiguration of human nature were "not accomplished in Christ from the start, from the moment in which he assumed our nature?" (Palamas, 1983, p. 76). Maximos Aghiorgoussis is even more explicit, writing:

> Everything that Christ did throughout his earthly life was based upon the assumption that humanity was already saved and deified, from the very moment of his conception in the womb of Mary. (Aghiorgoussis, 1992, p. 41)

A. N. Williams uses similar language when describing the effect of the "sheer joining of divinity and humanity in a single person", likewise making explicit reference to Jesus' conception in the womb. "From this first moment," Williams writes, the moment when humanity and divinity were united and "newly conceived in a woman's womb", salvation occurs (Williams, 2007, p. 38).

This does not mean that the events of Christ's life were completely irrelevant, but it does mean that the crucifixion (as does every other moment in Christ's life) points back and makes clear what Christ did by assuming a creaturely nature. For Palamas, Aghiorgoussis and Williams it is the union of the divine and the created in the person of Christ that is efficacious. While they do not use the language employed in this book, it could be said that for them, divine activity is due to the *relationship* of created and divine natures in Christ, rather than any particular *activity* that Christ does.

This view has been called a "physical view", which "derives the deification of the human race from its hypostatic union with the incarnate logos of God" (Mantzaridis, 1984, p. 29). Moreover, according to Mantzaridis, it was held by Irenaeus, Athanasius and Gregory of Nyssa (Mantzaridis, 1984, p. 29), all of whom are crucial for the development of the Eastern Christological paradigm. According to them, it was the "very assumption" of human nature by Christ that "had the effect of healing and transforming it" (Gendle, 1983, p. 140, n. 41). It is "physical", because the important part is the *contact* between the divine and the created natures in the person of Christ. It is the coming into contact with the divine that deifies the created. Again, it is not something that Christ or God does after this contact; it is simply being in contact with the divine that is efficacious and deifying. Thus, Gerald O'Collins notes of Gregory of Nyssa that the deification of creation is "rooted" in the fact that "Christ entered into a kind of physical contact with the whole human race" (O'Collins, 1995, p. 298).

So important is the emphasis on the contact, or "sheer joining", between the created and divine natures in the person of Christ that some commentators even claim that the deification that Christ affects is "automatic" (Kelly, 1965, p. 377) or "mechanical" (Finch, 2006a, p. 107). George Maloney, too, writing about Athanasius' theology, claims that Athanasius often "gives the impression that Christ has automatically redeemed us" (Maloney, 1968, p. 138). Deification is not something that requires a human response, nor does God differentiate between those who are the recipients of God's creating and deifying. The deification of creatures extends automatically and mechanically to all creatures, by virtue of sharing the same nature as Christ.

Perhaps the most important exponent of this view is Gregory of Nazianzus, whose letters against the Apollinarians have produced arguably the most important and simple Christological formula. Gregory writes:

> For that which is not assumed he has not healed; but that which is united to his Godhead is also saved. If only half Adam fell, then that which Christ assumes and saves may be half also. (Gregory of Nazianzus, 1954, p. 218)

The immediate context of the passage is the argument over whether Christ had a human mind or not, yet it can be inferred that Gregory makes this point because of the role of contact. Gregory's contribution that if Christ did not have a human mind then the human mind is not saved, demonstrates the centrality of contact and union for Christology. Divine activity/influence is achieved by coming into contact with creation. Whatever comes into contact with the divine cannot help but become deified.

In another letter Gregory solidifies this approach by continuing that if the Apollinarians could believe that God could "save man even apart from [assuming a] mind" then it would be possible for him to save humanity "also apart from the flesh [and] by a mere act of will" (Gregory of Nazianzus, 1954, p. 221). In other words, Gregory is arguing, if God did not have to become incarnate, then he would not have done so, and the reason that God had to become incarnate is because it is the contact that is crucial. The argument over the assumption of a mind demonstrates this; if Christ did not assume a human mind, then the human mind was not saved. It is the union between the two natures in Christ that is efficacious.

—

Again, it is not just the Eastern Fathers who expounded this view, despite the fact that it is characteristic of the Eastern Christology; some Western Fathers also pointed to the contact or union of the created and divine in Christ as being efficacious. Hilary of Poitiers, for example, writes that humanity is "sanctified by association with this mixture" (Hilary of Poitiers, 1982, p. 53). Lewis Ayres also notes that the "word's union with humanity in the person of Christ [as] the means of salvation" becomes increasingly important for Augustine (Ayres, 2004, p. 308). Even Teilhard de Chardin, in the twentieth century, acknowledged that "beatification coincides with a certain degree of physical incorporation in the created being of Christ" (Teilhard de Chardin, 1969, p. 167) and that deification is a state of "permanent Eucharistic union" (Teilhard de Chardin, 1969, p. 16), achieved through physical contact with the body of Christ in the Eucharist (Teilhard de Chardin, 1969, pp. 17–18).

a.1. *Communicatio idiomatum*

It is easy to see the connection between the Eastern understanding of the achievement of deification through the assumption of the created nature by the second person of the Trinity (i.e. the incarnation) and the doctrine of *communicatio idiomatum* that was outlined in the previous chapter. Both *communicatio idiomatum* and the "divine mechanism" of contact or "sheer joining" point to the union of divine and created natures in the single subject and person of Christ as the basis of divine influence. Or, perhaps more specifically, the divine mechanism of "sheer joining" demonstrates how it is that the union of natures in Christ can be the way that the human Jesus can perform the eternal divine activity of formal causality, conservation, ontological grounding, and definition of creation.

The conclusion of the previous chapter, therefore, regarding the union of divine and created in Christ as the basis of a theory of divine activity, is further supported by the theology of the incarnation in the Eastern Christology.

b. The divine mechanism and creation

The contact, "sheer joining", or union that the Eastern Christology emphasizes as the divine influence is the act of deification. The act of creation is completed in the deification of humanity, and this is achieved through the assumption of created nature by the divine person of Christ. However, the complete rejection of temporal categories means that God cannot do something new; God cannot create and then deify, any more than God can create and then save.

Likewise, deification cannot simply be the end of one activity that continues or perdures through time, as this also places temporal categories onto divine activity. God only does one thing, once. This means, as it has already been concluded, that creation and deification are identical: there is no difference between them.

As it has been argued in Chapter 3, "Divine Activity", there cannot be a clear demarcation (at least not from the divine point of view) between creation, conservation and deification. Creation and deification, for God, are not two distinct acts but are exactly the same. This does not mean that creation and deification are simply two ends of the same process (there cannot be any succession to divine activity); it means that creation and deification are both literally the same thing, which God does once, eternally. It is only from the subjective perspective of temporality that this single act is interpreted differently. There is one grace, eternally emitted by Christ, and this one grace is responsible for the conservation of the world, which, when interpreted differently, looks like creation or deification, depending on the perspective (see Hume, 2002, p. 57; Williams, 1999, p. 86). If Christ is responsible for deification, then this must be identical to the act of creation; the event of the incarnation must be responsible for creation.

Indeed, Edward Oakes comments that the Bible teaches that "the act of creation is intimately bound up with Christ's own incarnation" (Oakes, 2016, pp. 4–5). God only creates through the word, and the word only influences the world through a union with created nature (i.e. *communicatio idiomatum*). What this means is that the contact between the created and the divine in the person of Christ is the act of creation; the incarnation *is* the "initial" act of creation. The incarnation must already be a creative act (i.e. the individual human Jesus is created in the incarnation, otherwise adoptionism is unavoidable); however, because this creative act is responsible for creating the "firstborn of all creation" (Colossians 1:15; although not firstborn in the Arian sense—Christ is divine), so this creative act is responsible for creating the whole universe.

Again, it is worth making clear that this single act is not an efficient act: it is not an immediate and direct bringing into being and completing. The eternal divine act eternally defines what creation is, and this allows the universe to spontaneously self-create. The incarnation does not immediately and directly provoke or cause the universe to come into existence with the Big Bang (the incarnation is not co-terminus with the Big Bang); rather, the relationship between divinity and created nature in the single person of Christ is the eternal defining and holding creation in being. Christ's assumption of created nature is the formal cause of creatures, whose mutability means they can efficiently create themselves.

—

This raises an obvious problem. The question of how the incarnation (which happens in first-century Palestine) can affect an event that precedes it by fourteen billion years is easily solved by emphasizing that the divine event is *not* an efficient cause but a formal one. The incarnation, therefore, does not need to precede the Big Bang in order to account for the Big Bang's being able to spontaneously cause itself out of nothing; this is not how divine activity and influence function (i.e. divine activity is not univocal with, or in competition with, material cause and effect [i.e. created activity]). As it has already been claimed, there is not a connection between the eternity of divine activity and the beginning of the universe.

However, this does not solve the problem of how the incarnation, the coming together in the person of Christ of divine and creaturely natures, can happen without there being a nature *to begin with*. In other words, temporal dimensions may not necessarily present a problem, but the incarnation assumes that creaturely nature *already* exists in order for Christ to be able to assume it, and if creaturely nature already exists before the incarnation, then the incarnation cannot be responsible for its creation. How is it that the incarnation, which is a "sheer joining" of two disparate natures, is responsible for the creation of one of those natures (indirectly and mediately by defining the mutability that allows for that nature to self-create)?

The answer to this question is to look at the idea of assumption as a "subtraction" rather than an "addition"—an interpretation of the incarnation that is suggested by the theory of *tzimtzum* in the Kabbalah, a school of Jewish mysticism.

The incarnation as *tzimtzum*

The theory of *tzimtzum* is a theological idea that explains how God created the world by emptying himself and creating a "space" in which creation can happen. Hans Jonas, a Jewish philosopher, describes *tzimtzum* as follows:

> *Tzimtzum* means contraction, withdrawal, self-limitation. To make room for the world, the *En-Sof* (infinite; literally, No-End) of the beginning had to contract himself so that, vacated by him, empty space could expand outside of him: the "nothing" in which and from which God could then create the world. Without this retreat into himself, there could be no "other" outside God, and only his continued holding-himself-in preserves the finite things from losing their separate being again into the divine "all in all".
> (Jonas, 1996, p. 142)

Creation happens through God "first" contracting and withdrawing God's infinite nothingness to leave a space that is not God in which God could create. Yet, importantly, this space is not a physical space in the scientific sense—there is no such thing as empty space (Einstein, 1954, p. vi); it is a nothingness out of which space can spontaneously erupt (see Krauss, 2012, p. 170). This eruption, it has already been noted, is not provoked by God directly and immediately, but by the sheer fact that it is *not* God and, as "not God", must be mutable and, as mutable, able to self-create: "The world has to have aspects of non-necessity or contingency in order to be a world at all" (Haught, 2000, p. 40). *Tzimtzum*, as a doctrine of creation, claims that God creates by self-emptying.

—

There are other Christian theologians who have recognized the importance of *tzimtzum* for Christian theology, despite it being a Jewish doctrine. However, neither Moltmann (2001, pp. 145–6), Polkinghorne (1988, p. 61), Peacocke (2001, p. 87), nor Fiddes (2001, p. 185) notice the similarities between *tzimtzum* and the Christian doctrine of *kenosis*, which is the doctrine that Christ achieved the incarnation through self-emptying. The *tzimtzum* that Jonas describes can be seen as identical to the *kenosis* that Paul describes; both describe a creation through divine self-emptying. One describes the "limited" creation of a single human being (i.e. *kenosis*), while the other describes the "general" creation of the whole universe (i.e. *tzimtzum*).

Some theologians have hinted at such a comparison. Celia Deane-Drummond, for example, notes that "there are similarities between the incarnation and the creation of the world", pointing specifically to the "condescension" that characterizes both acts. However, she stops short of identifying them as the *same* act, indicating that the incarnation is God doing what God did in creation to a deeper degree (Deane-Drummond, 2009, pp. 114–15). Keith Ward, too, acknowledges that the incarnation is the "very same non-temporal act by which God also creates and consummates the created order" (Ward, 2001, p. 152). But, again, he sees those two events as the same "kind" or "type" of event, without seeing them as identical.

By understanding Christ as the sole and unique agent of creation, and by rejecting "multiple" graces (so that the "only channel for the outpouring towards us of sanctifying grace" [Meyendorff, 1964, p. 231] is likewise the only channel for the outpouring of creating grace), this book argues that these two self-emptyings are completely identical; they are precisely the same act. *Tzimtzum* and *kenosis* are not God doing the same thing twice; they are literally the same act. *Tzimtzum* just shows how the *kenosis* of the incarnation can be the *creatio ex nihio*. The incarnation is the "initial" act of creation, not because the incarnation efficiently causes the Big Bang (this would turn the "initial" act of creation into a secondary cause, and thus univocal with, and in competition with, created activity) but because the incarnation formally causes the ontological conditions that allow for the Big Bang to spontaneously happen—"nothing . . . *is* unstable" (Krauss, 2012, p. 170).

The creation of the world, which is the creation of an ontological (rather than material) space, out of which the universe can spontaneously self-create by virtue of being mutable (i.e. not God), is achieved through the self-emptying of the second person of the Trinity in the incarnation. The incarnation is not a means to an end—Christ does not become a creature *and then* create—the incarnation *is* creation. The incarnation must be creative in a very limited sense (i.e. that a human is created by it), but it is now argued that the incarnation is creative in a wider and more general sense (i.e. that the whole universe is created by it).

This provokes two important implications: (a) that the assumption is not the "addition" of something that already exists, but the creation of something (the universe) out of nothing (God), achieved through an "emptying" or "subtraction", and (b) therefore, the ontological space that God creates through self-emptying (out of which the universe spontaneously self-creates) *is* the human Jesus; the human Jesus is the definition of what it means to be created and is the ground of being for the whole universe—the universe is Jesus-shaped.

a. Incarnation as "subtraction"

The *locus classicus* of the Christian doctrine of *kenosis* is found in the letter of Paul to the Philippians, in which he writes:

> Though he was in the form of God, [Christ] did not regard equality with God as something to be exploited, but emptied himself, taking the form of a slave, being born in human likeness. And being found in human form, he humbled himself and became obedient to the point of death—even death on a cross. (Philippians 2:6–8)

What is important here is the absence of an explicit reference to the incarnation being an "addition"—that Christ, in becoming human, takes on something additional. Paul seems to imply that "taking the form of a slave" is achieved, not by taking on something, but by emptying himself. The implication is that the emptying itself achieves the incarnation, rather than the incarnation being the addition of something more or extra. Extrapolating this implication, it could be argued that the human nature is "left behind" or is the result of the emptying of the divine.

David Brown, in his book *The Divine Trinity*, seems to argue for a similar position, writing that *kenosis* "reduc[es]" the "divine reality" to a "human nature, initially no more than a foetus" (Brown, 1985, p. 231). The created nature of Christ is the result of a "reduction" of the "divine reality". Hilary of Poitiers also seems to imply this when he writes that God "contract[s] himself even to conception" (Hilary of Poitiers, 1982, p. 50).

The incarnation, then, is not the addition of something into the person of Christ. The incarnation is the creation of something "new", and this something "new" (i.e. created nature) is the result of the restriction of the divine in the person of Christ: it is the ontological space "left behind". The created nature (i.e. mutability) is what is created when the divine (i.e. immutability) is "removed". However, the incarnation is an addition in the sense that "after" the self-emptying, the person of Christ "contains" two natures. What it means is that the second nature is caused by, and results from, the retraction of the first. Thus, the self-emptying is not a complete abandonment of his divinity. As Hilary continues, this "contraction" of the human nature in the person of Christ means that the divine person of Christ becomes human "without departing from the power of God" (Hilary of Poitiers, 1982, p. 50).

There are some who criticize *kenosis* on the assumption that it means that "God became man and subsequently became God again" (Brown, 1985, pp. 102–3; see also Evans, 2002, p. 254); however, this is a misunderstanding of what *kenosis* is. Thomas Torrance, like Hilary, counters this criticism by making it clear that Christ "did not abandon his own immortality" (Torrance, 1978, p. 82). In fact, Michael Gorman writes, many of the Fathers "were very concerned" that Christ's self-emptying was "not the termination of his deity" (Gorman, 2009, p. 28). Augustine, for example, argues that Christ did not "[lose] the form of God" in taking the "form of a servant" (Augustine, 1991a, p. 77).

One of the reasons why Christ could not have emptied himself of all divinity is that this implies that the divine is measurable. The eternal, rather than infinite,[26] nature of God means that God cannot be measured or divided. If God were infinite, God could be continually and endlessly divided; however, as God is eternal, God cannot be divided at all—God is eternally simple. To put this same point differently, to claim that Christ could retract or empty himself of all of divinity would imply that the divine is exhaustible (a point that is wholly rejected by Gregory of Nyssa's doctrine of *epekstasis*, or "eternal progress").

Not only is it impossible for Christ to empty himself of the divine, it is crucial that he does not. The incarnation is, to borrow the definition of *tzimtzum*, the creation of an ontological space that eternally defines what creation is and keeps it in being, i.e. conservation. The universe no longer needs an "act of creation" (understood as an efficient act of God at the beginning of time that brings the universe into being); all the universe needs is "not to be God", and the mutability of "not being God" makes the spontaneous eruption of time, space, and all matter inevitable. However, that "space" still needs to be "in contact" with God in order to keep it in being. Both the rejection of the ability of the immaterial to influence matter (as suggested indirectly by neo-Darwinism) and the divine "mechanism" of the Eastern Fathers are adamant that the divine and the created have to be united in the person of Christ. If Christ was not God (at least not whilst he was human), then the connection between God and creatures in the person of Christ could not happen. If this could not happen, then God could not influence the world and therefore creation could never have happened.

The divine self-emptying, therefore, is the creation of a created nature that, by definition, must be in contact with the divine nature, as this retraction happens in the single subject of Jesus Christ. The self-emptying, therefore, *is* the sheer joining and contact between two natures. For the Fathers, this contact was the coming together of two previously existing natures; however, now, this contact is the creation of one nature by the retraction of the other.

a.1. Incarnation as "subtraction" and the analogy of being

One criticism of this idea could be that if the created nature is only in Christ as a result of the retraction of the divine, this implies that the divine and the created are in competition and, thus, are univocal, rather than equivocal. Sarah Coakley writes that a common misconception is to confuse "natures as interchangeable contraries" with "differing items of the same order", thus implying that the created and divine natures are "competing against one another for the same space" in the person of Christ (Coakley, 2002, p. 147). However, this is not the case. Christ is not more human the less divine he is. Just as the divine cannot be exhausted, so the emptying is not a literal emptying, as if Jesus is less God than before.

A solution to this problem can be found in the doctrine of analogy of being. In the previous chapter it was argued that the full union of the natures in the person of Christ meant that whilst they are still not univocal, they could not be entirely equivocal either. This is, to a large extent, the implication of the *communicatio idiomatum*, which claims that the one person of Christ has both a divine energy and a created energy, both of which are manifested in the actions of the single subject. The natures outside of Christ are equivocal—that has been established in Chapter 3, "Divine Activity". However, inside of Christ's person, the unity of the natures means that they are no longer considered equivocal. This means that there can be some interaction or influence between them (or at least from the divine to the created).

To put this a different way: the retraction is in the *person* of Christ, not the divine *nature*. Thus, Anselm claims that Christ did not assume human form so that "the divine and human natures are one and the same, but in such a way that the divine and human person are one and the same"

(Anselm of Canterbury, 1998, p. 250). Gerald O'Collins, too, notes that the unity "exists on the level of person", whereas "the duality [exists] on the level of natures" (O'Collins, 1995, p. 192).

The unity between the two natures is not on the level of nature. However, because there is a unity of natures on the level of personhood—a unique unity that occurs nowhere else and can occur nowhere else—there is not a complete equivocality between the two natures. Outside of Christ's person there can be no interaction and no contact between God and creatures; within the divine person of Christ there is an interaction and a contact. However, this does not mean that the natures within Christ are univocal; the causal relationship must not be considered as efficient just because the relationship between them is no longer equivocal. Just as with God's creation of the universe, so it is with the creation of the created nature in Christ (which is now the same event, i.e. the creation of the created nature in Christ *is* the creation of the universe): God's activity does not conform to the "cause and effect" relationship that characterizes the universe. The two natures are not in competition with each other, but there is a direct and immediate relationship and influence between them. There is something that is midway between this efficient and formal causality, and this something uniquely happens, eternally, in Christ.

The analogy of being means that the two natures are not in competition with each other for the same "space" within the person of Christ, i.e. univocal. Christ is not part God and part creature; Christ is both fully human and fully divine. However, at the same time, the complete and full union of those natures within the single person means that they are not equivocal; there is some influence between the two of them, which allows for the self-emptying of Christ to be, analogically, the cause of the created nature. That cause is the divine and the created nature coming into contact, which allows for a *communicatio idiomatum* and, therefore, allows the transcendent and equivocal God to influence (conserve) the material universe.

b. Jesus as ontological space
The incarnation is not coterminous with the Big Bang, which would make the divine act efficient or secondary causality. The incarnation creates an ontological space out of which the universe can spontaneously create.

This ontological space is the human being of Jesus of Nazareth, whose life as the firstborn of creation defines what it means to be a creature and, through that definition, grounds the universe in being and conserves it. As Ian Barbour notes, without God's eternal (*not* "continual") conservation "the universe would collapse into nothingness" (Barbour, 1971, p. 31). Basil Hume, too, writes that the existence of life is "sustained by a constant willing by God" (Hume, 2002, pp. 189–90). This conservation is the contact between the divine and created in the person of Christ. The human Jesus is the ontological space that conserves the universe.

However, an important distinction needs to be made in order to avoid falling into an incorrect understanding of what is being claimed here. The space that is created by the *tzimtzum* is an ontological space, not a material space. This means that the universe is not contained *in* Christ but is defined by him. The ontological space that is created by the self-emptying is Jesus-shaped, and that shape defines and describes the universe, which spontaneously erupts from nothing, materially "outside" of Christ but ontologically "within" Christ—as Teilhard writes, "nothing, Lord Jesus, can subsist outside of your flesh" (Teilhard de Chardin, 1978, p. 132).

The claim that the ontological space that is responsible for the creation of the whole universe as being *in* the person of Christ and thus identical to the human Jesus of Nazareth is not to be confused with an appeal to pantheism. To assume that this idea demands pantheism is to confuse the efficient and formal causality that this book has constantly distinguished between. In this book, drawing on the eternity and simplicity of divine activity, I argue, instead, for deism, i.e. there is a single divine action, preserving the transcendence of the divine. This means that the space that emerges from the self-emptying (*tzimtzum/kenosis*) of God in the second person of the Trinity is not a material space; the space that God creates in the person of Christ is not identical to the material universe. As Torrance notes, God (or the person of Christ) is not a "receptacle" in which creation moves (Torrance, 1978, p. 38). The human being that is created by the self-emptying is a *part* of this universe, not coterminous with it, yet as the firstborn of creation (in terms of importance rather than temporality) defines that creation.

Once that creation has been defined then it can spontaneously self-create. Thus, as Paul Davies writes, "the possibility immediately arises of space and time ... popping into existence, without the need for prior causation" (Davies, 2013, p. 50). Stephen Hawking, too, notes that "because there is a law like gravity, the universe can and will create itself from nothing" (Hawking, 2010, p. 180). More explicitly, Lawrence Krauss claims simply that "nothing ... *is* unstable" (Krauss, 2012, p. 170).

The neo-Darwinian synthesis sufficiently explains how something that is unstable can produce anything and everything, and the application of this as a universal ontology easily explains how the universe can emerge from nothing. The reason, therefore, why there is something rather than nothing is because that nothing is unstable; however, the reason that nothing is unstable is because of the incarnation, which creates an ontological "nothingness" that is not God, and because it is not God, it is mutable and passible. The "popping" into existence from nothing fourteen billion years ago is made possible, because God, through Jesus in the incarnation, defines what creatures are by emptying himself of immutability; Jesus eternally (i.e. non-temporally) causes the Big Bang.

Implications for the pre-existence and humanity of Christ

The idea that the incarnation as an event is identified with the eternal divine act of creation (which is conservation) has further implications that it is important to point to here. The first of these is that it means that if Christ conserves creatures (i.e. keeps them in being) by becoming created, then the incarnation itself must be an eternal event. More specifically, it denies that there is a "pre-existent" Christ, or, as Thomas Morris writes, that there is "a time before the Son began to exemplify human nature, a time at which he was not a man and yet existed" (Morris, 1986, p. 41).

Secondly, drawing on the fact that the evolution of humanity is an entirely accidental occurrence, and that there is nothing unique or special about humanity such that there cannot be a definite demarcation of humanity from other creatures, then Christ's humanity must likewise be accidental. To put this another way: to claim that Christ became *human* is unacceptably exclusive; rather, Christ became a creature and therefore his incarnation is relevant for all creatures.

a. The pre-existence of Christ

The incarnation, identified as the unique and single divine activity that is responsible for the conservation of the universe by defining and grounding it, precisely because it is divine activity, must be eternal. This means that there cannot be a time before the Son was human because this means that God and divine activity are subject to succession. In other words, the word of God, who is the unique and sole agent of divine activity, the one mediator between God and creation, is always the *incarnate* word.

Some theologians find it difficult to separate the pre-existent Christ from the human Jesus. Henri de Lubac, for example, writes that "Christ existing before all things cannot be separated from Christ born of the woman, who died and rose again" (de Lubac, 1947, p. 174). Indeed, de Lubac also argues that the "grace of Jesus Christ" can never be separated "from the incarnation itself" (de Lubac, 1969, p. 81). What Christ does (i.e. the bestowal of grace), which includes creating, Christ does through the incarnation. Other theologians, who also agree with this connection, see this as being demanded by the first chapter of Colossians, in which Paul writes that "all things have been created by him and in him". Christopher Mooney, for example, notes that "apparently" Paul is always thinking of "the concrete, historical God–man" and "never the word independent of his humanity" (Mooney, 1966, p. 170). Richard Kropf, too, notes that Origen and Hippolytus, "in line with the prologue of St. John's Gospel", have a tendency to relate these passages in Paul to the "word made man" (Kropf, 1980, pp. 144–5). Christ creates through becoming created, and this becoming created—the formal and ontological causation of the universe, a relationship of dependency of one nature on the other in the person of Christ—is an eternal event; there is not a "time" when God, in Christ, is not self-emptying.

This does not mean that the human being of Jesus Christ is eternal. It has already been argued that the fact that creatures receive their being from eternity does not mean that creatures are themselves eternal, and the same must apply to the human being of Christ; claiming that the divine person eternally self-empties does not mean that the human being of Christ does not begin to exist at a certain point—eternity does not mean infinite time. As Brian Leftow puts it, "if the Son is timeless and incarnate, it does not follow that his human nature is co-eternal with his divine nature" (Leftow, 2002, p. 299). This means it can be said that there was a time when Jesus was not born; but this does not mean that there was a time when the divine person was not self-emptying. The two claims mean different things.

b. Christ as creature
The chapter on neo-Darwinism and the subsequent chapter on the anthropological implications of neo-Darwinism made one important point: due to the primacy of preservation and the power of accumulation, there cannot be a clear demarcation between humanity (or any other creature for that matter) and the rest of creation. Humanity is properly a part of the universe from which it came and cannot be separated ontologically from it. The difference between humanity and the rest of the universe is quantitative (i.e. by degree), not qualitative (i.e. by kind). As Darwin himself put it:

> Now all these modified descendants form a single species, [and] are represented as related in blood or descent to the same degree. (Darwin, 2009, p. 369)

Or, as Richard Dawkins says, making the same point, there is "only a smeary continuum" and the species is "an arbitrary stretch of a continuous flowing river, with no particular reason to draw lines delimiting its beginning and end" (Dawkins, 1986, p. 264). This means that there cannot be anything said about humanity that does not also include and refer to other creatures, which must include the incarnation. If there is nothing to separate humanity from the rest of creation, then there is nothing to say that the incarnation is not relevant to other creatures. More importantly, if the evolution of humanity is only an accidental occurrence, then the humanity of Christ must likewise be entirely accidental. Christ does not need to become *human* in order to achieve the grounding and defining (i.e. conservation) of the universe; Christ needs to become *created*.

Gregory of Nazianzus' famous and beautiful Christological formula, "that which he has not assumed he has not healed", can help to clarify this point. If Christ deifies (which must now be identical with creating and conserving) by coming into contact with that nature (i.e. the assumption of it), and there is nothing that separates human nature from other created natures (if indeed the concept of nature has any meaning after neo-Darwinism), then Christ must come into contact with—i.e. assume—not just a human nature in the incarnation, but the whole universe.

To put this rather crudely: if Christ is a male human, then Christ is only relevant to male humans, and only male humans are created. If Christ is a human—his masculinity being accidental and irrelevant—then Christ is only relevant to humanity and only humans are created. However, if Christ is a creature—his humanity being just as accidental and irrelevant (as neo-Darwinism argues) as his masculinity—then Christ is relevant to all creatures. Christ could have been a mouse, a cicada fly, a tree, or a proton, and nothing would have changed: the universe would still be here in all its splendour.

However, claiming that Christ must have assumed the whole universe in the incarnation does not mean that the whole material universe is the created body of Christ, as has already been argued. The space that Christ assumes (i.e. creates) in the incarnation is an ontological space that is not coterminous with the universe, rather than a material space that "contains" the universe.

The claim that Christ assumed the whole universe does not mean that the particular first-century Palestinian male human existence of Christ is not real. Robinson puts this point thusly:

> To be a "universal man" is not to have every human quality, but to be the sort of person of whom we recognize *in* the individual that which transcends the individual. We see in him what *each* of us could be—in his own unique way. What attracts and judges us is not the man who has everything—that merely oppresses us—but the man in whom we can glimpse a vision of the essential.
> (Robinson, 1973, p. 73)

It does not matter, therefore, whether Christ was Jew or Gentile, whether he was right- or left-handed, or whether he was male or female, or human or mouse. What matters is that in this person, there is something universal that all creatures recognize.[27] Frances Young makes the same observation:

> However complex historical reconstruction may be, there must have been something about Jesus which elicited the response whereby each follower saw him as the answer to his deepest needs and claimed to see God disclosed in him. (Young, 1977, p. 39)

In much the same way that it was argued that the single, eternal divine activity is interpreted differently depending on the subjective need of the individual (see Aquinas, 1998, pp. 294–5), so all creatures can see and find in Jesus Christ the "answer to their deepest needs", and the ground of their very being, despite the fact that he was "accidentally" (i.e. not necessarily) manifested as a human.

Conclusion

This chapter has argued that the incarnation, interpreted through the lens of *tzimtzum*, can be the single divine activity. Building on the conclusion of the previous chapter, which saw the *communicatio idiomatum* (the relationship between the divine and created natures in Christ) as the way that God influences the world, this chapter has argued that the incarnation is the way that this *communicatio idiomatum* comes about. This is further supported by the divine "mechanism" that characterized the Eastern Christology—it is through the sheer joining of God and creatures in Christ that this divine activity is achieved.

Recognizing the problem with seeing the coming together of two natures as being the causation of one of them, the incarnation must be seen as a "subtraction", rather than an "addition". This does not mean that this subtraction is not creative—there are two natures in Christ after the subtraction—but it does mean that Christ does not assume something that has already been created; the assumption *is* the creating.

However, this assumption is not an efficient cause—i.e. materially causing the Big Bang—it is a formal cause. The creaturely body of Jesus is the ontological space that is left behind after the subtraction of divine nature in the person of Christ. This ontological space "defines" what it means to be created—i.e. mutable—and, therefore, "after" this incarnation, the universe can spontaneously create itself from nothing, because nothing is inherently unstable.

Thus, the incarnation in first-century Palestine can be the divine act of creation (a) because the *kenosis*—which is the *tzimtzum* outlined in the Kabbalah—brings the created nature into being through self-emptying, and (b) because that emptying is formal, not efficient, and so does not need to happen "before" the Big Bang. It is *ex nihilo* not *post nihilo*; there is no identity between the act of creation and the first moment of creation.

CHAPTER 6

Participation, Imitation, and Neo-Darwinism

It is necessary to pause for a moment to review what I have already argued for. Firstly, neo-Darwinism, by providing evidence of an absence of divine influence on the world, helps to clarify and nuance a theological idea that can be traced back to the Early Church: namely, that the transcendence of God means that God cannot influence the universe (which includes the act of creation itself) without becoming created. This points to the role of Christ, who is the unique and sole agent of divine activity and the one mediator between God and creation.

The relationship of divinity and creation in the person of Christ, therefore, *is* the relationship of dependence (i.e. conservation) that gives being to, and grounds the being of, the universe, which is the single, eternal divine activity. This relationship—which draws on *communicatio idiomatum* and what has been described as the "divine mechanism" of the Eastern Christology (i.e. that it is contact with the divine that is effective)—is the result of the incarnation. Drawing on *tzimtzum* as an interpretive category, the incarnation describes how the self-emptying *kenosis* eternally creates an ontological space—which is the human body of Jesus—that defines the universe as mutable and allows for the spontaneous self-creation of the universe. This nuancing of themes that are *already* present in theology—i.e. the transcendence of God, the non-temporal nature of divine activity (i.e. divine activity as relationship rather than act), and the role of Christ as creator—is the result of a sober conversation with the neo-Darwinian synthesis.

—

However, there is still a question that remains unanswered. If in Chapter 3, "Divine Activity", I completely reject immaterial influence (i.e. direct and immediate divine activity), and in Chapter 4, "The Person of Christ", deny that the universe is coterminous with the body of Christ, then how is it that this divine influence, which is confined to the person of Christ, is felt by, or extends to, all other creatures? Put slightly differently, if the relationship between divinity and creation is found *only* in the person of Christ, how is the result of that dependency experienced by, and how does it influence, all other creatures?

If, in Chapter 3, I argued for a single, non-temporal, and non-spatial (i.e. eternal) divine influence—which must be understood as a relationship of dependence and conservation between God and creation, rather than an "event" (i.e. cause and effect)—and in the two chapters on Christology I argued that the incarnation can be seen to describe how that relationship comes about and how the immaterial divine can influence the material creation, then in this chapter I explore what the relationship itself looks like.

The answer to that question, I will argue in this chapter, can be found in the doctrine of participation. Participation of the creature in the life of God has been a core Christian doctrine since the Early Church. Drawing on biblical ideas, the Church Fathers understood the life of the creature to be a sharing of the life of God. The creature only exists in and through the divine life of God. This idea can also be found in the Scholastic period and right through into contemporary theology. The persistent use of participation as a way of describing the being of the creature means that participation becomes the content of the doctrine of creation; to be created means to participate in the life of God. Participation in the life of God describes a relationship of dependency that understands God to be the sole and unique reason for the life of the creature, whose continuing life depends upon a "continuing" relationship with God.

It will also be argued that the doctrine of participation functions in a significantly similar way to the doctrine of imitation. Whilst the doctrine of participation is a crucial one that seeks to describe the content of the doctrine of creation, it makes little sense in a materialist context that is sceptical of immaterial influence. To put this same point differently: what does it mean to participate in the life of God and what does that look like?

The answer is imitation. It will be shown that imitation of Christ, itself an important and influential element of Christian doctrine, describes the relationship between God and creature in a way that is so similar to that of participation that they can be considered for all practical purposes to be synonyms; to participate in God is to imitate Christ.

However, what is even more important than the observation that participation and imitation are identical descriptions of the same relationship, is that this imitation can then find support in the neo-Darwinian synthesis. The ontology that it was argued must be the conclusion of the neo-Darwinian synthesis is, again, strikingly similar to the ontology that is argued for by participation and imitation. In fact, some biologists even use the language of imitation to describe the relationship between genes.

This means that, while in previous chapters in this book I have attempted to show how Christian doctrine must be nuanced in order to accommodate the neo-Darwinian synthesis, on the subject of what it means to be created (i.e. the content of the doctrine of creation and the subject of this chapter), there is remarkable agreement. How the Christian understands what it means for the creature to exist is the same as how the neo-Darwinist understands existence.

Participation

The idea that creatures participate in God is a way of describing how creatures are created. Vladimir Lossky, for example, writes that creation "implies the idea of participation in the divine being" (Lossky, 1957, p. 118). To be created by God is to participate in the life and being of God; conservation and participation are the same act seen from different perspectives.

Participation was an immensely important category for the Early Church. For the Greek Fathers, it was, if not a synonym, at least an image of deification (see Williams, 1999, pp. 123–4). Drawing on the language of 2 Peter, in which the author writes that humanity "may escape from the corruption" of the world and "may become participants in the divine nature" (2 Peter 1:4b), the early Fathers saw the end of human life as consisting of a close relationship with God in which they could enjoy the benefits of the divine life. In fact, according to Norman Russell, Cyril of Alexandria replaces the language of deification with the language of participation (Russell, 2004, p. 192).

Irenaeus, too, "constantly affirms" that salvation entails "participation in God" (Finch, 2006, p. 93). Likewise, in Origen's writing, "participation is the means by which deification is effected" (Russell, 2004, p. 154). Participation, then, is the way that the Patristic writers understood the end of creaturely life in deification. The end goal of the human life was to participate in, and share, the divine life; whilst God possessed that life by necessity, humanity possessed it by grace.

However, due to the fact that deification was the end of the whole process of creation (see Louth, 2007, p. 34), participation can be used as a description of creation as much as it can of deification. Deification may be participation in the life of God to a much deeper degree, but participation can be descriptive of the "natural" life of creatures as well. Participation does not just describe the creature's deification; it also describes how the creature can exist at all—"human life depends on participation in God" (Osborn, 2001, p. 230).

Pointing to this use of participation as a description of how creatures are created and how they are deified, Russell notes that participation has a "twofold aspect", describing both how creatures are "raised from non-existence to createdness" and how creatures "advance from createdness to transcendence" (Russell, 2004, p. 191). Thus, it is "in him we live and move and have our being" (Acts 17:28). It is in God that creatures have their existence; without the eternal formal act of God, creatures could not exist: "Without their participation in the word, the entire universe would revert to the nothingness from which it came" (Powell, 2003, p. 18; see Hume, 2002, pp. 189–90 and Barbour, 1971, p. 31).

Put simply, according to Cyril of Alexandria, "participation is the key to the understanding of our relationship with God" (Russell, 2004, p. 193). The relationship that is the eternal divine activity is participation.

—

As much as the doctrine of participation was an important category for the early Fathers and their development of the doctrine of deification, it was also an important category for the Scholastic theologians of the Middle Ages. A. N. Williams, for example, notes that Aquinas' doctrine of participation "radically exceeds classic notions of theosis in its scope", because Aquinas recognized that participation in the divine life must be a doctrine of creation as much as deification (Williams, 1999, pp. 65–6). Williams claims that Aquinas reached this position by disagreeing with Augustine. Whereas Augustine saw the need for grace as a direct result of the Fall (see Cary, 2000, p. 68), Aquinas argued that even without the Fall, the creature would still need grace (Williams, 1999, p. 79).

Grace, which is "not just pardon for the poor sinner" but "participation in the divine nature" (Rahner, 1966, p. 177; see also Meyendorff, 1964, p. 121), then, is just as descriptive of creation as it is of deification. The difference that this book has argued is that, whereas Aquinas and others would see creation and deification as two ends of a process of deeper and closer communion (i.e. the reception of *more* grace), this book argues that they are precisely *identical*, separated only by the subjective individual: "Grace cannot differ in degree [objectively]", because God treats all creatures the same, "but grace may differ in degree from the perspective of the subject" (Williams, 1999, p. 86). This is a position made necessary by the eternity and simplicity of divine activity.

a. Participation as dependence

The doctrine of participation in the being of God is essentially a description of how creatures are entirely dependent on God for their being. Participation, by describing how it is that creatures are made in the image of God, must describe a relationship between creatures and God, and that relationship, this book has consistently maintained, must be one of dependence—the dependence of creatures on God. Or, to put it another way, participation "expresses a derivative mode of being" (Russell, 2004, p. 194): God has being "essentially, [yet] we have it derivatively and by participation" (Keating, 2007, p. 97).

Daniel Keating, in his book on deification, draws attention to this aspect of participation. Participation is a relationship between two—the participator and the participated—but these two are always unequal;

participation always describes a relationship in which one is entirely dependent on the other. Keating writes:

> Participation necessarily requires a relation between two things that are unequal ... [an] unequal relationship between what is essential and what is derivative. (Keating, 2007, pp. 97–8)

The necessary dependence of the one on the other creates an unequal relationship. God has being by necessity, and creatures have it contingently and derivatively.

This relationship of dependence is the same as that which I described in Chapter 2, "Theological Anthropology", in relation to a neo-Darwinian theory of original sin. The Church sided with Augustine, not (at least not primarily) because it agreed with the anthropology and cosmology that he built up around that theory, but because it agreed with his assertion that the creature was completely and utterly dependent on God. This relationship of dependence now has a specific metaphysical framework: participation.

a.1. Participation as subjective

Not only does participation denote a relationship of dependence, it also denotes a relationship that is permanent. Despite the fact that participation as deification entails a "dynamic [relationship] in which we advance from createdness to transcendence" (Russell, 2004, p. 191), the relationship between God and creatures is always between two unequals. If the relationship changed, then creatures would no longer be dependent on God, and, thus, "there would be no relationship of participation" (Russell, 2004, p. 148). Regardless of how dynamic participation is, if it entails the creature becoming identical to God, then the creature can no longer participate in God. Participation, by its very nature, must be permanent. Keating observes that "if something simply *becomes* another thing, then it can no longer be said to participate in that thing", thus "participation entails (and guarantees) both a true relation and a true distinction" (Keating, 2007, p. 98). In this way:

> [Participation] enables us to grasp how we are genuinely related to God and can partake of his life, without jeopardizing the infinite distance that distinguishes the uncreated Trinity from all creatures. (Keating, 2007, p. 103)

The permanence of the unequal relationship of participation means that, if the creature, through deification, becomes more divine, then it must always be possible for the creature to become more divine. Or, as Keating puts it, regardless of how much creatures participate in the divine life as the basis of their own existence, there is always a greater distinction than relation.

Gregory of Nyssa termed this greater distinction *epekstasis*, or "eternal progress"; in other words, the creature's journey to deification and communion with God can never finish. Regardless of how close to God the individual comes, there is always a greater distance; there is an infinite distance between God and creatures. Gregory of Nyssa says of this "perpetual progress" (as Meyendorff calls it [Meyendorff, 1978, p. xvi]) that "the soul rises ever higher and will always make its flight higher" (Gregory of Nyssa, 1978, p. 113) and that "[the soul] makes its way upwards without ceasing" (Gregory of Nyssa, 1978, p. 113). Gregory of Nyssa continues that:

> For this reason we also say that the great Moses, as he was becoming ever greater, at no time stopped in his ascent, nor did he set a limit for himself in his upward course. Once having set foot on the ladder which God set up (as Jacob says), he continually climbed to the step above and never ceased to rise higher, because he always found a step higher than the one he had attained (Gregory of Nyssa, 1978, pp. 113–14)

There is no limit in the ascent to God because the "increasing desire" for God is never satisfied (Gregory of Nyssa, 1978, p. 116). The infinite and eternal God can never be approached. There is, instead, an "incessant transformation into the likeness of God" (i.e. imitation) and an "ever-greater participation in God" (Malherbe and Ferguson, 1978, p. 12). God is infinitely transcendent of creation and therefore greater participation in the divine life is always possible. Deification, then, becomes the description of an (endless) process, rather than a state; i.e. the deified creature is the creature that is continually becoming more like God, not the creature that is like God.

Gregory of Nyssa is not the only exponent of this view. Henri de Lubac, for example, notes that Pseudo-Dionysius, Maximus the Confessor, and John of Damascus also held a doctrine of the perpetual progress of the individual into the divine life (de Lubac, 1998, p. 43). Gregory Palamas also understood the relationship between the creature and God in this way:

> Clearly it will develop infinitely ... the saints, communing in the grace of God and rendered through that communion more and more able to contain the divine radiance, will receive grace upon grace from God himself, its infinite and unfailing source.
> (Gregory Palamas, quoted in McDaniel, 1992, p. 82)

God is the infinite source of grace and the more that communion with God is achieved, the more one can commune to a greater degree.

—

The "eternal progress" of participation in the divine life means, importantly, that "proximity" to God must be subjective: if there is always an infinite distance between the creature and God, then the proximity of the individual to God must be subjective. If there is not an infinite distance between the creature and God (i.e. not an ever-greater possibility of participation in the divine life), then God must be quantifiable and measurable—something that God's eternity (i.e. non-spatiality) vehemently denies.

This is the same as the point that there can only ever be a subjective distinction between creation and deification, due to the eternal (i.e. singular) nature of divine activity. Creation and deification are both participation in the divine life, and it is only the subject who distinguishes between them. As Williams, Hume, and Scuka have already been quoted as saying, "sanctifying grace cannot differ" from the perspective of God, but it can differ "from the perspective of the subject" (Williams, 1999, p. 86); God cannot bestow more grace—God eternally bestows grace—but the creature can be influenced by it differently. (This is the same point that Augustine and Aquinas make, that there is only one divine activity from the divine perspective but many activities from the created perspective.) The creature already participates in the life of Christ, and deification becomes the subjective realization that the creature *already* participates in Christ. Gregory Collins seems to agree with this understanding:

> We are already the children of God (1 John 3:2), but we often do not remember it. The spiritual life means appropriating in one's subjective experience what God has already granted through the objective grace of baptism. (Collins, 2010, p. 226; see also de Lubac, 1947, p. 39 and Scuka, 1989, p. 86)

The infinite distance between God and the creature means that the creature cannot objectively become deified—as if the distance between God and the creature can be measured and come to an end—and this also means that the creature cannot become "more" deified or become "closer" to God than any other creature. There is just as great a distance between God and humanity as there is between God and, say, a mouse; to claim that God is closer to humanity is to quantify God and claim God is measurable, i.e. that God and creatures are univocal. This means, therefore, that neo-Darwinism is correct in its claim that creatures and humanity must be considered equal.

If there is no objective way to claim that humanity is closer to God than a mouse, then participation in God must be subjective and so what it means to participate in God is different for different creatures. As Samuel Powell writes, "the diversity of creatures implies that there are various ways in which creatures can participate in God", so that "each species participates in God in a way that distinguishes it from other species" (Powell, 2003, pp. 48–9). All creatures participate in God, but this participation is not the same for all creatures. Although, as much as it is not the same, one participation is not better or closer to God than the other.

Thomas Merton also seems to hint at this idea when he writes:

> A tree gives glory to God by being a tree. For in being what God means it to be it is obeying him. It "consents", so to speak, to his creative love. It is expressing an idea which is in God and which is not distinct from the essence of God, and therefore a tree imitates God by being a tree. (Merton, 1961, p. 30)

A tree participates in God (as do all creatures) just like humans do, but *how* trees participate in God and what it means for a tree to participate in God is different for all creatures.

Any criticism of this idea (such as Rolston's claim that "trees do nothing voluntary" [Rolston, 2001, p. 62]) is unacceptably anthropocentric. As it has already been claimed, the idea that self-consciousness (or even "simply" consciousness) represents an improvement upon other, non-conscious, life is vehemently rejected by neo-Darwinism. The freedom that is supposedly required to respond to the grace of God is only a subjectively different way of responding to God; it is unacceptable to claim that a tree cannot participate in the divine life because it is not free (especially when it is not entirely clear that humanity is free anyway!). There is nothing about the way that humanity participates in God that makes it closer to God than other creatures; proximity to God—both in terms of the distinction between creation and deification and in terms of relation of other, non-human creatures and God—is always a subjective category.

b. Participation as Christological category

The doctrine of participation led Aquinas to the doctrine of the analogy of being. As Samuel Powell writes, it is "because creatures have their being by participation" that there is an "analogical relation between them and God" (Powell, 2003, p. 48). This book, however, has argued that the analogy of being must now be considered a uniquely Christological category. This move from "philosophy to Christology" is further supported by (and in many ways dependent upon) the fact that the doctrine of creation has likewise been moved from a general theological doctrine to a specifically Christological one.

Christ is the unique agent of divine activity precisely because the relationship between God and creation happens in his person. If participation is what it means to be created in the image and likeness of God, then participation *must* be a uniquely Christological category. Creatures do not participate in God's life; creatures participate in Christ's life. Grace, in which all creatures participate, is the image of God that is the ontological space created through the incarnation and is identical with the human body of Jesus.

The Christological dimension of participation has been recognized by other commentators: Simo Peura, for example, notices that "participation in Christ is participation in divine nature" (Peura, quoted in Saarinen, 1997, p. 74). Participation in Christ is to be in communion with the divine nature; or, to put this same point differently: the person of Christ is the one mediator between God and creation. Tuomo Mannermaa, the founder of the Finnish Lutheran School that rediscovered the doctrine of deification in the theology of Luther, of which Peura was a member, also notes the Christological dimension of participation. He writes that "participation in the divine life of Christ is at the core of the doctrine of divinization" (Mannermaa, 2005, p. 2).

To be deified, therefore, is to participate in the divine life *in* Christ; there is no relationship between God and creatures outside of Christ. However, F. W. Norris, expressing sentiments that seem to lend direct support to the theology that this book has espoused, specifically links participation in the divine life with the incarnation:

> Participation in God's nature became possible only through the incarnation of the son . . . [Christ's] humanity is the link to our participation in God. (Norris, F., 1996, p. 420)

Christ is the only mediator between God and creatures and in his person is the only relationship between divine and created natures; therefore, participation (i.e. being created) *is* only possible through the incarnation, and Christ's humanity *is* the only route of creatures to God (and the only route of God to creatures). Russell also explicitly points to the incarnation as the means by which the creature participates in God, claiming that participation in the divine nature "means that the Son communicates himself" (Russell, 2004, pp. 181–2). In fact, "this dynamic participation in the Logos is only possible because of the incarnation" (Russell, 2004, p. 182). Quoting Cyril of Alexandria, Russell continues that:

> For "the divine nature" [of 2 Peter 1:4] is God the Word together with the flesh. And we are his "offspring" even though he is God by nature, on account of his taking the same flesh as ourselves... we are united to the Father through him as through a mediator. For Christ is, so to speak, a frontier between supreme divinity and humanity. (Russell, 2004, p. 201)

The reference to *communicatio idiomatum* serves to support the interpretation that this book has offered. Christ, who is the "frontier between divinity and humanity" and "mediator" between God and creatures, is the person in whom creatures participate as the ground of their being. The relationship of divinity and creation in the person of Christ is responsible for creatures' participation in the divine life, which is the content of the doctrine of creation (i.e. conservation).

This Christological locus of the doctrine of participation can also be inferred from biblical theology. Powell, for example, notes that the concept of participation is "found mainly in Paul's notion of being in Christ" (Powell, 2003, p. 45). This is clearly a reference to Paul's letter to the Colossians, which has already been quoted several times in this book. In the opening chapter, Paul writes:

> [Christ] is the firstborn of all creation; for in him all things in heaven and on earth were created, things visible and invisible... all things have been created through him and for him... and in him all things hold together. (Colossians 1:15ff.)

While Paul does not use the language of participation (in fact, aside from the single reference found in 2 Peter, there is no reference to participation in the Bible), there is clearly support for the idea that participation in the person of Christ is the content of the doctrine of creation (i.e. conservation); to be created is to participate in Christ, in whom "all things hold together".

c. Participation as ontology

All of this points to the fact that participation is a Christian doctrine of ontology. Participation in the life of Christ is what it means to be created; it is what it means "to be". This ontology, because it is entirely dependent on the grace of God, describes a permanent relationship between two unequals—one entirely dependent on the other. This participation, subjectively different for each creature (which agrees with neo-Darwinism), is descriptive of what it means for creatures to be created and what it means for creatures to be deified; it is only the individual creature who, realizing that their life is already a participation in the divine life of Christ, can subjectively differentiate between creation and deification.

This subjective tension between creation and deification is demanded by the fact that the equivocality of divinity and creation means that God and creation are infinitely separated; there is an infinite gulf between divinity and creatures that is only bridged by Christ. Regardless of how close to God the creature gets, there is always a greater and infinite distance between them, which also means that there is a profound equality between all creatures; all creatures receive the fullness of grace and all creatures are equally close to God—it is only those who respond to that grace differently that can be subjectively closer (i.e. deification).

Imitation

"Participation expresses a relationship which is metaphysical, not corporeal" (Russell, 2004, p. 147). However, if participation does not express a corporeal relationship, then it cannot describe a relationship in a paradigm that rejects immaterial influence (and Platonic forms), such as this book has argued for. This does not dismiss metaphysics as a legitimate pursuit, but it recognizes that if God cannot "immaterially" influence the world then an "incorporeal participation" must be rejected by the conclusions to Chapter 3, "Divine Activity". Participation must describe a physical or corporeal relationship to make it credible for the theology that this book has argued for. Put slightly differently, the same point can be made by asking the question: what does participating in the person of Christ look like? The answer is imitation: to participate in the divine life through Christ is to imitate Christ.

The link between participation and imitation is not a novel observation. Montagnes, for example, notes that "one cannot separate participation from imitation", writing that "God himself is the form [in which] the created being participates by imitation" (Montagnes, 2004, pp. 35-6). Kathryn Tanner, too, seems to imply that participation and imitation are closely related (if not synonyms), writing that "human beings are images of God by participation" (Tanner, 2011, p. 71). The explicit link of the idea of participation with image could be interpreted as a link between imitation and participation (drawing on the common root of the words "image" and "imitation" [see Augustine, 1991a, p. 283, n. 35]).

However, where Tanner might be vague in her linking of imitation and participation, Susan Wood is clear: "*imitatio Christi* [is] based on the fact that the Christian participates in the life of Christ" (Wood, 1998, p. 141). For Wood, the creature imitates Christ *precisely* because creatures participate in the divine life of Christ. Wood stops short of claiming that participation and imitation are synonyms, but there is clearly a relationship between the two; imitation of Christ is because of (and thus the sign of) participation. Norman Russell, too, explicitly understands imitation and participation to have a close relationship, also linking these with the analogy of being. Russell writes that "analogy, imitation, and participation thus form a continuum rather than express radically different kinds of relationship" (Russell, 2004, p. 2; see also Ayres, 2004, pp. 321ff.). Again, Russell stops short of arguing for synonymity, but this is not ruled out either.

Drawing on this link between imitation and participation, this section will argue that the idea of imitation holds a strikingly similar role in the theology of the Scholastics and, particularly, in Patristic theology. Likewise, it will show that the "metaphysics" of imitation are identical to those of participation in that both describe identically the relationship between God and creation. It is this identicalness in the description of the relationship between God and creation that cements the claim that imitation and participation can be understood as synonyms. Thus it supports the claim that imitation is at least the outward manifestation of participation, if not able to completely replace the doctrine of participation in a Darwinian paradigm that rejects an immaterial influence and seeks only material explanations.

a. Imitation and deification

In the same way that participation is used as an image of deification and the creature's relationship with God, so imitation is used as a way to describe what deification means to the creature.

The implications of the eternity of divine activity have already been keenly emphasized. If imitation is an image of what it means for the creature to be deified, then it must also be an image of what it means for the creature to be created; as Hume and Aquinas have already been quoted as supporting, it is only the creature that can distinguish between the two. The very same point was made in relation to participation. Any image that is used as a description of deification must also be used as a description of creation. Thus, the very ground of being of all creatures is imitation of Christ. Aquinas points to the use of imitation as a synonym for creation when he writes that "the creature exists only to the degree that it descends from the primary being, and it is called being only because it imitates the first being" (Aquinas, 1998, p. 58).

In fact, the use of imitation as a description of what it means to be created is in many ways a response to, and permitted by, the opening chapters of Genesis, which claim that humanity (which must now include *all* creatures) is created in the image of God. Anthony Hoekema makes a similar connection between "image" and "imitate", noting that the concept of "image of God" should be used as a verb as well as a noun: creatures are not just an image of God but have *to image* God (Hoekema, 1994, p. 52). To be made in the image and likeness of God, therefore, at least for Hoekema, is nothing but the idea that imitation of God is the basis and content of the doctrine of creation; to be created is to imitate God.

b. The metaphysics of imitation

Perhaps more important than imitation being used as an image for creation and deification in addition to participation (and thus suggesting a similarity between the two) is that they both describe the relationship between God and creatures using the same language.

Elizabeth Castelli, in her book *Imitating Paul,* discusses the metaphysical implications of imitation. She writes:

> The notion of imitation presupposes at least two important and related things: a relationship between at least two elements and, within that relationship, the progressive movement of one of those elements to become similar to or the same as the other ... imitation is then the celebration of identity, in the sense that sameness implies the quality of identicalness ... it is the struggle to write the identity of the model onto the copy. (Castelli, 1991, pp. 21–2)

Imitation, then, is not just a moral principle in that the correct thing to do in any given situation is to do as Christ did. There is a much deeper and more central and fundamental element to imitation in that there is a struggle towards ontological identity. However, importantly, this struggle can never be overcome. Castelli continues that "mimesis is always articulated as a hierarchical relationship, whereby the 'copy' is but a derivation of the 'model' and cannot aspire to the privileged status of the 'model'" (Castelli, 1991, p. 16), as a result of which "there exists in the notion of imitation this tension between drive to sameness and the inability to achieve it, an inability which creates a hierarchy" (Castelli, 1991, p. 75). Within this hierarchy, "the model is imbued with perfection and wholeness, the copy represents an attempt to reclaim that perfection" and, therefore, "mimesis becomes a derivative function, in that it attempts to reproduce an unattainable origin" (Castelli, 1991, p. 86).

As for the previous quotation, it is the copy trying to become the model that represents the tension between the two. This, as Castelli notes, means that the copy derives its being from the model: it is the model that is responsible for the existence of the copy. Precisely because the copy is a copy of the model, the model is responsible for the being of the copy. Imitation, then, like participation, is a description of how the creature derives its being from God, who is entirely responsible for it, and on which the creature wholly depends.

This derivation between the two means that there is always a difference between them. No matter how closely the creature imitates Christ, there is always a difference. If there were not a permanent difference between the two, the identity of the copy would be utterly destroyed; the creature would literally become Christ. As Augustine writes, "if an image perfectly matches that of which it is the image, it is coequated with that, not with the image" (Augustine, 1991a, p. 215). This permanent tension means that it is always possible to imitate Christ more; it is always possible to become closer to God. This in turn means that there is an "infinite gulf" between creature and God (O'Collins, 2002, p. 26), an infinite gulf that characterizes Gregory of Nyssa's doctrine of *epekstasis*.

This "infinite gulf", others have commented, is the basis of the doctrine of the analogy of being, which "sets a 'certain likeness' within a 'greater unlikeness'" (Loughlin, 1999, p. 144). Montagnes, for example, writes:

> The analogical unity that unites them consists in the fact that creatures imitate God to the extent that they can, to the extent that their nature permits, without attaining the fullness of the divine perfection. (Montagnes, 2004, p. 36)

The doctrine of the analogy of being, as it has already been noted, states that there is a likeness to God (univocality) and a greater unlikeness (equivocality). The likeness of the creature to God is always because Christ, who is creature, is both divine and created (not because there is a general likeness between God and creatures; analogy of being is a Christological, not a philosophical, doctrine).

However, imitation is as much a description of unlikeness as it is likeness; imitation describes how the creature can be like and unlike God at the same time. This state of likeness and unlikeness is identical to, or synonymous with, the way that participation understands the relationship between God and creatures. More importantly, it is a relationship that is channelled through Christ, which means that it is always an imitation of *Christ*. In the same way that participation must be a Christological category, because it is *in* and *through* Christ that the creature is created, so imitation must be a Christological category. Imitation is specifically an imitation of Christ.

What is most important about this exposition of the metaphysical or ontological implications of imitation is that it is almost identical to the way that participation outlines the same metaphysical implications. Not just this, but the language that was used to describe the participation of the creature in the divine life of God is precisely that which is used to describe imitation. The creature "derives" its being from God, who is the ground of being. Likewise, this "derivation" produces a tension that means there is always an "unequal" relationship. There is an attempt to imitate (or participate in) God through Christ as much as possible, but an inability to do so.

Imitation, then, as a description of the relationship between God and creatures, must be seen as being the *content* of the eternal divine activity. Causing the creature to imitate Christ *is* the content of God's divine act of creation. God does not, as it has been emphasized at length, bring the universe into being in an efficient sense, but in a formal sense. To put this differently: God does not create and then cause creatures to be like God (i.e. creation *then* deification); being like God *is* the act of creation (i.e. creation *and* deification as conservation).

c. Imitation and hierarchy

In the previous section it was mentioned in passing that imitation functions as a hierarchy. The permanent and necessary tension that exists between the imitator and the imitated means that there is always an ontological hierarchy between the two. However, this hierarchy means that imitation of Christ is not a *direct* and/or *immediate* imitation of Christ: it is an *indirect* and *mediated* imitation. Castelli continues her exposition of imitation by writing:

> Christ is to Paul as Paul is to the Corinthians; Paul asks for an act of *imitatio Pauli*, which mirrors his own *imitatio Christi* ... [imitation, therefore] presupposes a hierarchical structure: community/Paul/Christ/God. (Castelli, 1991, p. 112)

This is clearly an allusion to Paul's claim in his first letter to the Corinthians, in which he exhorts the Corinthians to "be imitators of me, as I am of Christ" (1 Corinthians 11:1). As much as Paul urges the community to imitate Christ, that imitation is always *through* him. This is a very interesting and important point. Imitation might very well be an imitation of Christ, but it is never a *direct* imitation of Christ (apart from the twelve apostles).

This link between hierarchy and imitation is especially present in the theology of Pseudo-Dionysius, whose comments on the nature of God have already been noted in Chapter 3, "Divine Activity". In the *Celestial Hierarchy*, one of the very few surviving works penned by him, he writes that "the goal of a hierarchy, then, is to enable beings to be as like as possible to God and to be at one with him" (Pseudo-Dionysius, 1987, p. 154) and that "those closer to God should be the initiators of those less close by guiding them to the divine access, enlightenment, and communion" (Pseudo-Dionysius, 1987, p. 158), thus "each designation of the beings far superior to us indicate ways in which God is imitated and conformed to" (Pseudo-Dionysius, 1987, p. 166) and "hence on every level, predecessor hands on to successor whatever of the divine light he has received and this, in providential proportion, is spread out to every being" (Pseudo-Dionysius, 1987, p. 178).

The divine influence, which is the ground of being (i.e. participation), is mediated "to all other beings, including ourselves" by others *through* imitation (Pseudo-Dionysius, 1987, p. 178). Grace—the image of God—which is the ground of being, is not immediately passed on by God, but indirectly, and mediated through imitation.

Importantly, as Pseudo-Dionysius is keen to point out, "God himself is really the source of illumination for those who are illuminated, for he is truly and really Light itself" (Pseudo-Dionysius, 1987, p. 178). Just because those further down the hierarchy are "illuminated" (i.e. influenced by God) by others, it does not mean that the influence is not completely and utterly divine.

As participation and imitation are identical, so this hierarchy can also be applied to participation. Creatures participate in God as the source of their being, but that participation is *through* other creatures. William Cavanaugh writes, "the members of the Body [of Christ] are not simply members individually of Christ the head, but cohere to each other as in a natural body" (Cavanaugh, 1999, p. 184). More importantly, "the members . . . participate in each other" (Cavanaugh, 1999, p. 184) and this "participation in one another [is] through our creation in the image of God" (Cavanaugh, 1999, p. 192).

Of course, the immediate context of this quotation is the union of all creatures in Christ; the creature does not participate in the being of Christ without also being united to all other creatures. However, there is also here a hint of what this chapter has argued for.

There is a biblical precedent to this idea. The author(s) of the book of Genesis write:

> When God created humankind, he made them in the likeness of God. Male and female he created them, and he blessed them and named them "Humankind" when they were created. When Adam had lived for one hundred and thirty years, he became the father of a son in his likeness, according to his image, and named him Seth. (Genesis 5:1b–3)

Seth, according to the author of Genesis, is made in the image and likeness of *Adam*, not *God*. This, of course, cannot possibly mean that Adam is responsible for Seth in the same way that God is responsible for Seth; nor could it mean that Adam is the ground of Seth's being in the same way that God is the ground of Seth's being, but it does mean that Seth imitates and participates in God *through* Adam. Seth imitates Adam who imitates God; it is the image of God, but mediated through Adam, as Hoekema writes: "if Adam was still the image-bearer of God, as we saw, we may infer that Seth, his son, was also an image-bearer of God" (Hoekema, 1994, p. 15). To put this same point differently, God is still the *formal* cause of Seth, but Adam is the *efficient* cause of Seth; God is still primarily responsible for Seth, and Seth is completely dependent on God for his being, but the image in which Seth is created is mediated by Adam.

Fran O'Rourke makes a similar point, writing that:

> As each being partakes—in its own measure—in divine perfection, it also shares in a parallel manner in God's creative activity, but is itself continually sustained and maintained by God's presence. (O'Rourke, 1992, pp. 12–13)

To claim that divine activity is mediated through the hierarchy does not claim that there are other "ultimate" sources of creation beside God; it simply claims that the grace of God (which is the image that all imitate) is passed on to others "down" through the hierarchy.

Importantly, this does not mean that Adam has more grace than Seth because Adam passed that grace onto Seth; the infinity of grace (see Hume, 2002, p. 57) and the simplicity and indivisible or unquantifiable nature of grace means that it is impossible for anyone to have more grace than anyone else; it is the response to that grace, or allowing that grace to influence them, that is crucial, not the amount of grace that one has (see Williams, 1999, p. 86).

c.1. Hierarchy and subjectivity

This indirect and mediated imitation of Christ means that the same must be said of participation and, therefore, the ground of being. The creature does not derive its being directly from God and Christ, but from others. This is the content of Pseudo-Dionysius' theology, whose doctrine of the hierarchy was not so much a description of creatures' movement to, and proximity to, God as a way of describing how the divine influence was "passed down" to other creatures (see Nelstrop, 2009, p. 109).

However, if the ontological "gap" between God and creatures is infinite, this does not mean that the imitated creature (i.e. the creature that "passes on" the imitation) is closer to God. Such a claim would suggest that God is quantifiable and measurable and has already been criticized in relation to participation. Not only this but, as Hume notes, that grace is infinite (Hume, 2002, p. 57), so grace can never be diluted or lessened as it is passed on. Regardless of how many times grace is "passed on", it is always infinite; it is impossible to be "closer" to God, so it is impossible to "have more" grace—the individual only responds to that grace differently. In this direction, O'Rourke writes:

> The outpouring of the divine gift is infinite in its source; it is not diminished, however much it is shared . . . God's plentitude is infinite and never decreases in itself. Nor is God's power divided in causing different and distinct things; he causes all by

a single power and the diffusion of his goods is not diminished.
(O'Rourke, 1992, pp. 240-1)

When creatures pass on the grace of God (which is the image in which they were created and which they imitate) they do not pass on less than what they receive: the grace of God, like the divine nature and the divine activity, is simple and cannot be divided or exhausted.

However, while grace is not diminished, O'Rourke continues that "while God loves all things in a constant and single act of will, some are better because he wills more good to them, i.e. he loves them more" (O'Rourke, 1992, p. 261) and that "the gradation of perfection among creatures is rooted in their degrees of participation in and proximity to transcendent and absolute perfection" (O'Rourke, 1992, p. 263). This, it has been argued, cannot be the case. The eternity of God, made clear in Gregory of Nyssa's doctrine of *epekstasis*, means that proximity to God is impossible; there is also room for greater participation and imitation. All are created equal, all pass on the divine grace "down" the hierarchy through imitation, but none are closer to God. This is also affirmed by neo-Darwinism, with its characteristic denial of ontological distinction among creatures, or what O'Rourke calls the "*saltum qualitatis*" (O'Rourke, 1992, p. 267)—the qualitative jump. God loves all creatures and treats all creatures the same, and all receive the same grace; it is the response to that grace that differs from creature to creature, not God's bestowal of grace.

Likewise, in the same way that the tension between imitation as creation and imitation as deification is entirely subjective—God eternally, singularly, and simply causes creatures to imitate him, and it is the individual creature who sees that single event as multiple—and so proximity to God (i.e. deification) must be entirely subjective, so what it means to imitate God is also entirely subjective.

It is important to note that some commentators also criticize this anthropocentric claim that humanity imitates Christ "better". Linda Zagzebski, for example, writes:

> There are many ways to imitate Christ, and that is why there are so many different kinds of saints ... which saints a person

should imitate is probably determined partly by his or her life circumstances and partly by his or her individual moral philosophy. (Zagzebski, 2002, pp. 328–9; see also Meister Eckhart, quoted by McLeod Bryan, 1961, p. 65)

Whilst this almost certainly refers to differing personalities and life circumstances, there is no reason why it cannot be applied to all creatures. As it was argued in the previous chapter, the particular first-century male Jewishness of Jesus does not exclude women, Gentiles, or even non-humans from seeing in Christ the ground of their being and the source of their life. Each creature passes on the imitation of Christ (and in doing so causes others to participate in the divine life), but that does not mean that the imitated is closer to God, nor that there is an objective idea of what that imitation is.[28]

d. Imitation and participation

At every point, therefore, there is, at the very least, a connection or, at the very most, a synonymity between imitation and participation. For both the Church Fathers and the medieval Scholastics, there is an identity between imitation and participation as an image and description of deification and how the creature exists in the first place. More importantly, the relationship between God and creatures that imitation and participation describe is identical. Both understand the very being of the creature to be a sharing in the divine being who is responsible for the existence of the creature.

This sharing creates an unequal relationship, in that one has being necessarily and the other derivatively, and so depends on the other for its being. Yet this sharing of the divine being does not create a univocal relationship between the two. Rather, mediated through Christ, whose incarnation is necessary for the relationship, this relationship is analogical (which some have noted can be used as a synonym for both participation and imitation). This relationship, for both participation and imitation, is characterized by an attempt to write the identity of the model onto the copy but always, and by definition, being unable to do so. Importantly, it is always an imperfect imitation, or, crucially, an imperfect *replication*. This creates a "perpetual" movement of the creature towards God, who, being infinitely distant, creates a subjective tension between creation and deification (i.e. *epekstasis*).

Christian ontology and neo-Darwinism

If imitation and participation can be seen as synonyms based on the fact that they describe the relationship between God and creation (a relationship that in Chapter 3, "Divine Activity", I identified as the single, eternal divine activity) in identical language, then neo-Darwinism can be understood to contribute to this discussion by giving that description another focal point.

The ontology that it was claimed was suggested by the neo-Darwinian synthesis is mirrored in the ontology that the doctrine of participation and imitation describes. The ontology of imitation outlined above conforms in a significant way in that both use almost identical language to describe what it means to be created. It has already been remarked that neo-Darwinism is not a theory of how things come into being (i.e. a doctrine of creation), but describes how they are (i.e. an ontology), and that description has much in common with the description that Christian theology describes, as outlined above.

a. Neo-Darwinism and imitation

The use of the words "replication" and "copying" by biologists, when describing the neo-Darwinian synthesis, is immediately noteworthy. Not only can both words be used as synonyms for "imitation", but the word "copy" has also been used by theologians (as above) to describe the role of imitation. There is, therefore, immediately a significant connection between neo-Darwinism and imitation/participation. However, some biologists explicitly use the word "imitation" to describe the "process" of neo-Darwinian evolution.

While Darwin himself sparingly used the idea of imitation, making fleeting references to it in order to describe instinct and behaviour (Darwin, 2004, pp. 93–4), Richard Dawkins explicitly uses the word "imitation" to describe the "process" of genetic replication that is the core of evolution. In his book *The Selfish Gene*, Dawkins claims that the

DNA molecule, which is the prevailing replicating entity on this planet, "[is] a unit of *imitation*" (Dawkins, 2006, p. 192). DNA imitates itself, albeit imperfectly, in order to create something new. Elsewhere, Dawkins also argues:

> If individuals live in a social climate in which imitation is common, this corresponds to a cellular climate rich in enzymes for copying DNA. (Dawkins, 1999, p. 110)

The link between the "social climate" of imitation (which Darwin also alluded to) and the "process" of genetic replication is explicit. The "social imitation" (which can be understood to include the *imitatio Christi* explained above) is identical to the process of genetic replication that is responsible for all life.

In fact, as it has already been expounded, for Dawkins, the process of genetic replication (with its inevitable failure or openness to mistakes or errors) is only one specific example of a more universal or general principle, which, in his essay *Chinese Junk and Chinese Whispers*, he calls "universal Darwinism".

> The real unit of natural selection was any kind of *replicator*, any unit of which copies are made, with occasional errors, and with some influence or power over their own probability of replication. The genetic natural selection identified by Neo-Darwinism as the driving force of evolution on this planet [i.e. DNA] was only a special case of a more general process that I came to dub "universal Darwinism". (Dawkins, 2004, p. 149)

This universal Darwinism accompanied Dawkins' more controversial idea of memes, i.e. social replicators. Dawkins' idea was that ideas or concepts replicate and survive in exactly the same way that genes do. In *The Selfish Gene*, Dawkins writes that genes and memes propagate themselves "via a process which, in the broad sense, can be called imitation" (Dawkins, 2006, p. 192). Therefore, "imitation, in the broad sense, is how memes can replicate" (Dawkins, 2006, p. 194).

Despite the fact that the application of a biological idea to a social setting has been criticized (see McGrath, 2005, pp. 121ff. and Foster, 2009, pp. 102ff.), Dawkins' idea of the meme makes clear that the genetic process of replication is actually the manifestation of a more general and universal principle. This principle is exactly that which participation and imitation describe; what Dawkins calls "universal Darwinism", this chapter has called "participation". Not only is the use of language an important point of contact between these ideas, but the relationship they describe is also identical. For participation and imitation (which are identical and thus synonyms) the replication or copying of the model is always imperfect—there is always room or potential for closer imitation or greater participation—as perfect copying or replication would destroy the identity of the copy.

This imperfection is also a feature of neo-Darwinism, as Dobzhansky evidences when he writes that "every succeeding generation of a species resembles but is never a replica of the preceding generation" (Dobzhansky, 1982, p. 9). The imperfection that is responsible for the openness to failure that creates the diversity in the world is a direct result of the ontology of imitation that Christian theologians have described. For theology, it is important that the imitator/participator never imitates/participates in the imitated/participated perfectly because this would destroy the identity of the imitator on the one hand and compromise the transcendence of the imitated on the other. This means that a Christian ontology must always emphasize imperfect imitation and imperfect participation.

However, this is exactly the same for neo-Darwinism, which *relies* on the imperfect replication of genes (i.e. imperfect preservation) in order to account for the change and diversity in the world, which is subsequently selected through differential survival and reproduction rates. In other words, quite importantly, the neo-Darwinian synthesis can be explained by appeal to the theological ontology of participation. Genes are imperfect replicators, because such imperfect imitation/participation is a necessary feature of what it means to be created. The emphasis on accumulation and preservation in evolution is important to connect it with participation and imitation.

Evolution, then, is an inevitable part of creation, not because God designed the world to be that way, but because there is simply no other way to be created. "Once" Christ had become incarnate and, in his person, set up the relationship of dependence through the eternal emptying of divinity that created the ontological space that was the historical body of Christ, it was inevitable that the universe would spontaneously create itself and all the wondrous and diverse life would come into being. Evolution is not a process that is controlled or directed by God, but an inevitable consequence of the fact that to be created means to participate in and imitate the divine life (i.e. to be open to failure).

The ability of the universe to create itself as a result of the instability of "nothing" means that all God needs to do in order to ensure that the universe can come into being is to define the ontology that neo-Darwinism describes. This ontology is now described by imitation (and participation). Imitation of Christ is the relationship that is responsible for the neo-Darwinian synthesis, and ultimately why "nothing" is unstable and why the universe can create itself. This relationship is caused by the eternal coming together of created and divine natures in the person of Christ.

The claim that "nothing is unstable", which explains how it is that the universe can spontaneously self-create from nothing, is "simply" the application of the neo-Darwinian principle to physics. The "universal Darwinism" that Dawkins argues for, that the idea of imperfect replication can be applied outside of the narrow biological setting in which it was proposed, means that, ultimately, genetic mutation and quantum fluctuation are different manifestations of the same principle. Thus, "'natural selection' seems capable of application ... beyond the biological domain altogether, so as possibly to have relation to the stable equilibrium of the solar system itself, and even the whole sidereal universe" (Mivart, 1871, p. 22). This principle, it has been argued, is also described (albeit in theological/philosophical language rather than scientific) by participation/imitation. The incarnation and participation/imitation are the *same* "act" seen from the divine side and the created side respectively, therefore, the question of why is there something rather than nothing (i.e. creation) is answered by the ontology of imitation: there is something rather than nothing because the universe participates/imitates, which is the incarnation (remembering that the non-temporal nature of the divine act means that this does not happen *post-nihilo* [i.e. after nothing], but *ex-nihilo* [i.e. from nothing]).[29] The ontology of imitation that is caused by the incarnation, as this book has argued for, by agreeing with the neo-Darwinian synthesis, can explain why nothing is unstable and, therefore, how the universe can self-create.

b. Hierarchy and mediated divine influence

Another important similarity between neo-Darwinism and imitation is the rejection of an objective model and the subjective judgement of evolutionary success on the one hand or proximity to God on the other. The criticism of Gould's theory of "punctuated equilibrium"—that there is only a "smeary continuum" of life—focused on the fact that there is no such thing as a species of which each individual is a manifestation. Rather, the centrality of accumulation meant that the genetic replication is not the copying or imitating of a general and objective model but always the imperfect copying or imitation of the direct ancestor. There is nothing objective called a "mouse" that all mice imitate, there is just continual copying of a direct ancestor that means the descendants of a "mouse" will also look pretty much like a mouse but never identical to it.

The same phenomenon was argued to be a feature of imitation. The individual is called to imitate Christ—and the individual already imitates Christ—but this is not the direct and immediate imitation of Christ, but an imitation that is mediated through others; the individual is called to imitate Paul, who in turn imitates Christ. This led to the role of hierarchy, not as a way of demonstrating proximity to God (which can never be objective) but as a way of describing divine influence. Marilyn McCord Adams, for example, wrote:

> Just as God is the fontal source of natural being (*esse*) and goodness in creatures, so Aquinas envisions a cascading flow of grace: from Godhead into the human soul hypostatically united to it; from the soul of Christ into all the members of the body of which he is the head. (McCord Adams, 1999, p. 52; see also van Driel, 2006, p. 284).

The grace of God comes from the Godhead, through the created nature of Christ, and "down" into the rest of creation. The "rest of creation" is then called to "pass on" this grace; as Pseudo-Dionysius writes, "predecessor hands on to successor whatever of the divine light he has received" (Pseudo-Dionysius, 1987, p. 178). Hierarchy, then, is not a doctrine of how creatures approach God—the point is not to move "up" the hierarchy—but how grace reaches creatures. Thus, Nelstrop evidences that the function of hierarchy is "not that one should climb up to God, but that each should fulfil its role" (Nelstrop, 2009, p. 109).

This subjectivity is yet another important connection between the two ideas. Imitation, both in a theological context and a biological one, is never the imitation of an objective principle but the subjective imitation of a direct ancestor. In the same way that genes are "passed" on, or mediated, through different creatures, so the grace of God—which is the model/image of Christ—is passed on to each creature or mediated through other creatures. In this way, the two most important features of imitation/participation are shared by neo-Darwinism. Both participation and neo-Darwinism explain what it means "to be" as imperfect replication of a direct ancestor, with no objective model to copy from, and that imitation is always judged subjectively.

There is an important aspect of biological imitation that might be seen as a criticism for this approach to theological imitation. In his essay *Chinese Junk and Chinese Whispers*, Dawkins notes that "memes travel longitudinally down generations, but they travel horizontally too, like viruses in an epidemic" (Dawkins, 2004, p. 142). When this is taken in the context of this book, it could be argued that biological imitation, while it may share some important features with theological imitation, differs significantly from it, because it is restricted and limited, whereas theological imitation is much more open. Or, to put this differently: theological (or social) imitation is not always imitation of a direct ancestor but could be imitation of anyone.

However, while Dawkins is technically correct in making this point, he misses an important detail. Biological imitation is not different from sociological or theological imitation as he assumes; it is simply that the generations of the different imitations do not directly or perfectly match up. This means that memes will also "travel" (and subsequently mutate) longitudinally, but this longitudinal, or generational, "travel" does not match the same generations as genes. This gives the impression that memes travel horizontally (i.e. between two biological individuals of the same generation), yet that is only *biologically* horizontal: the meme has still "travelled" (and mutated) longitudinally from a social point of view. This means that there is not a difference between the way that biological imitation and sociological or theological imitation function, as Dawkins claims. It simply needs to be recognized that the generational "travelling" of memes and genes has no direct correlation.

Certainly, it must be maintained that biology and sociology or theology are not entirely identical. The scientific method, for example, is entirely inappropriate for theological study. In the same way, theological epistemology is different from scientific and biological epistemology. However, this does not mean that participation/imitation cannot describe an overarching ontology—a universal principle as Dawkins claims—of which neo-Darwinism is but a biological manifestation. Neo-Darwinism, therefore, is exactly what a theologian *expects* from a theory of evolution—not as a replacement for the Genesis narrative but explaining the genesis of all creatures through the principles of participation and imitation.

The incarnation is the doctrine of creation. Through the incarnation the second person of the Trinity creates an ontological space (identical with the creature Jesus) that defines what creation is (i.e. not God and therefore passible and mutable) and conserves it. This conservation, seen from the side of the creature, is participation, which is imperfect replication. Participation is the "universal Darwinism" of which biological evolution is just one manifestation. Evolutionary change—which is responsible for the "creation" of all "species"—is not guided, controlled or set up by God to create, but is the necessary consequence of a world that is mutable (i.e. open to failure), and this world is mutable because it is "not God", because Jesus achieved its creation through the emptying of God.

Conclusion

The previous three chapters can now be seen to ultimately provide an answer to the question of divine influence (or divine responsibility for creation) in a neo-Darwinian paradigm that rejects the ability of a non-material entity to influence the material universe.

Chapter 4 outlined that Christ must be the sole and unique mediator between creation and divinity and the sole and unique agent of divine activity and influence. This divine activity, being a relationship, is better understood as a relationship between creation and divinity, and this relationship is found in the person of Christ. Chapter 5 continued in this direction, showing how the incarnation is the reason for this relationship and outlining how it is that the incarnation, understood kenotically, already has the necessary tools to deal with being responsible for the creation of the whole universe. Using the Kabbalahic doctrine of *tzimtzum* as an interpretative lens, the incarnation becomes, not the *addition* of something extra, but the creation of something new through *subtraction*. The incarnation must already be a creative event in a limited sense (or adoptionism is unavoidable), but it must now be seen as a creative event in a much wider and universal sense.

Lastly, this chapter has shown that the doctrine of participation, identical to imitation, can explain how the incarnation can be a wider creative event that can influence each individual creature. Conservation, the eternal divine activity that is achieved by the union of divine and created natures in the second person of the Trinity through the incarnation, is *precisely* the same phenomenon as participation/imitation seen from the side of creation; participation is how Christ conserves the universe, and the universe can "spring" into existence by itself, because it participates. Using Pseudo-Dionysius' theory of hierarchy, the "contact" between the creature and God—a "contact" that the Eastern "divine mechanism" sees as being creative—is felt indirectly, mediated to the individual through others. This mediation spreads outwards through imitation from Christ to all creatures, both spatially to the furthest ends of the universe and temporally to the furthest ends of time (i.e. the first and last "moment" of time).[30] Creatures imitate (literally image)/participate in the image of God, which is the ontological space that is created by the *kenotic* assumption and is identical to the historical body of Jesus. This image is "passed on" through imitation so that all can be said to participate in the life of Christ and be conserved through "contact" with God.

This paradigm of participation and imitation, crucial as it is to complete this Christology and explain how doctrines of Christ solve the problem created by a neo-Darwinian theory of divine activity, actually agrees with the ontology that neo-Darwinism suggests. In this way, there is a significant agreement between theology and neo-Darwinism; original sin, participation, imitation, and neo-Darwinism all describe the same ontology—half-fixity, imperfect replication and openness to failure—albeit in different language.

God creates an ontological space (which is the formal cause of mutability) from which material creation (i.e. the universe) spontaneously self-creates. This ontological space is identical to the historical body of Jesus, which is the definition/ground/model of being. This makes the incarnation *the* divine act of creation. Participation/imitation is identical to this act and is the act of creation seen from the perspective of the creature; the divine act influences all creatures indirectly and mediately through participation/imitation. That participation/imitation is further identified with the neo-Darwinian synthesis (evolution), which does not describe how things come into being, but how things are, and which describes things coming into being as a result of their imperfect imitation/replication.[31] However, this conclusion doesn't take into account the central Christological event (at least in the Western Church): the cross.

CHAPTER 7

The Cross

With the conclusion of Part II, the question of what Christology looks like from a neo-Darwinian perspective has been answered: the transcendence of God has been emphasized, which places Christ in a much more central role than previously given. This is not a completely new approach; what I have done is draw on, and emphasize, elements that are already present in Patristic and Scholastic theology. By focusing on the role of Christ, I have shown how the incarnation can be the single, eternal act of creation and that the divine influence, mediated through Christ, is then further mediated through imitation to all creatures, thus showing how all creatures participate in Christ.

However, any discussion of Christology, and the role that Christ plays in the history of the world, can never be complete without a discussion of the crucifixion (see McGrath, 1987, p. 36). It is all very well, following the Eastern Christological paradigm, to emphasize the importance of the incarnation over and against the crucifixion, but to completely ignore it would be to overlook one of the most important events in the life of Christ.

Traditionally, the crucifixion's role is expounded exclusively in terms of salvation. The rejection of the paradigm of salvation, which is made necessary by evolution, seems to suggest that the crucifixion must be rejected along with it. Certainly, the Eastern Christological paradigm, as has already been noted, emphasized the incarnation over and against the crucifixion, but this does not mean that the crucifixion no longer has a part to play in Christian doctrine. This chapter will show that, not only can the crucifixion be expounded along un-salvific lines, but that these lines are, arguably, more fundamental.

So, the crucifixion should not be rejected; rather, it requires a different perspective. Quick, for example, notes that the difference between Johannine (i.e. Eastern) and Pauline (i.e. Western) theology is that the former sees the "incarnation [as] the starting point of theology [while] atonement is a derivative of this", whereas the latter understands it as the other way around: that "the incarnation is a means of [atonement]" (Quick, 1938, p. 189). This Johannine perspective will be taken in this chapter. The incarnation is not a means to the end of the cross—as if the only importance of the incarnation is to allow the cross to be possible—but, rather, the cross is an extension of, or "participates" in, the creating power of the incarnation. This connection between the incarnation and the cross will be explored below.

This chapter is organized into four sections. The first will consider those models of the crucifixion that understand it in terms of atonement. Drawing primarily on Gustav Aulén's influential division of crucifixion theologies into three distinct categories, this will delineate between those theologies that understand the cross as being an offering to the Father and those that see it as an offering to the devil.

The second section will outline the criticisms against these models of the crucifixion. Firstly, it will consider theological objections to the notion of salvation, questioning whether it can make sense for the cross to be an offering to the Father or the devil. Secondly, it will reconsider some of the reasons that evolution denies the paradigm of salvation, questioning whether there is anything from which creatures require salvation in the first place.

The third section will consider a few different models of the crucifixion that understand the cross outside of the paradigm of salvation. The most important of these is the role that evolution plays in understanding the cross. Teilhard de Chardin, for example, understands the cross as having creative, rather than salvific, influence. This shows that, not only can the cross survive outside of the salvific paradigm, it can also thrive.

The final section will develop this idea of the cross's relationship to creation. Understanding the cross as creative, it will show how this understanding builds on and develops the theology of the incarnation that this book has expounded.

The cross as atonement

It is common to understand the cross as being linked, if not exclusively then predominantly, with the idea of atonement. This does not mean, however, that there is only one way to understand that atonement. While there are many theories of atonement, most can be grouped into two essential models: those that understand the Father to be the object to whom the crucifixion is directed and those that understand the devil to be the object.

a. Crucifixion as satisfaction: sacrifice to the Father

One of the most important models of the cross as atonement is that which sees the cross as being a satisfaction for sin. Sometimes called the "Latin" theory, particularly by Gustav Aulén (whose book *Christus Victor* is a seminal exploration of theories of atonement), this sees the cross as a price paid to God in order to make amends for wrong done to God in the Fall.

The role of legal rhetoric and imagery is important for understanding the development of this theory. The overemphasis on legal rhetoric has already been explored in relation to original sin, and it is clear how this model of atonement relates to this overemphasis in Tertullian, Ambrose, and Augustine. However, the most crucial aspect of this model is the role of divine justice. Before everything else, God is supremely just—as God was for the Deuteronomist (see Harris, 2003, p. 160)—the good are rewarded and the bad are punished. Humanity, after the Fall supposedly recorded in Genesis, is deserving of punishment, and it must pay reparations to God in order to restore its standing in God's eyes.

The world, because it is created by God, is a rational and logical place. This means that injustice cannot exist, regardless of whether God desires otherwise. In the same way that God cannot create a squared circle or add two to two and get five, even if God wished to, so God cannot violate the divine justice that rules over the world. If God were to simply forgive the sin of humanity, without receiving reparation, then divine justice would be shattered; this idea of divine forgiveness, therefore, is completely out of the question. It is not a question of whether God *would* forgive, but whether God *could* forgive.

So central is this to the "Latin" theory of atonement that John Stott writes that the law "must take its course" and that even if "God loves us sinners and longs to save us", God "cannot do so without violating the law which has justly condemned [humanity]" (Stott, 1986, p. 115). The law plays such a central and even tyrannical role in this theory of atonement that divine love is only "allowed to operate" within the limits of divine justice. Even more so, Aulén claims "there is a suggestion that the idea of the divine love is regarded with some suspicion" (Aulén, 1931, p. 130). Interestingly, the constricting nature of divine justice must not always be seen as diluting or frustrating divine love. F. Noel Palmer, for example, argues that if God were to simply forgive us or "let us off", this "would only damn us more deeply" (Palmer, 1949, p. 20). For Palmer, then, the failure of God to simply forgive without reparation is not due to a limitation on God's part, but is a part of God's love of creatures.

Perhaps the most important adherent of this model is Anselm, "who centred on the divine honour rather than the divine love" (O'Collins, 1983, p. 149). The main features of Anselm's theory are as follows:

> On the one hand, [Anselm rejects] forgiveness of sin which would be a bare remission of penalty . . . [and] on the other, of an optimistic conception of man's capacity to perform all that was needed. (Aulén, 1931, p. 85)

Anselm, focusing on divine justice, rejects the idea that God would simply forgive humanity without it first meriting or deserving that forgiveness; yet, in the next breath, he denies that humanity, marred as it is by sin, could ever merit or deserve that forgiveness. Christ, then, becomes crucial, as

> [n]o one can pay [the recompense] except God, and no one ought to pay except man: it is necessary that a God-man should pay it. (Anselm, 1998, p. 320)

Christ is essential, because divine justice demands that humanity pays for its sin and Christ, being God, is the only human who can overcome the constraints of sin and pay that reparation. The death of Christ, then, is the reparation that is due to God because of human sin; without Christ, humanity would and could never repay God for the offence caused to God in the Fall.

b. Crucifixion as cosmic battle: ransom to the devil

The other major theory of the cross as atonement, which Aulén calls the "Classic" theory and attributes it to the New Testament, the Greek Fathers, and Martin Luther (Aulén, 1931, pp. 6–15), sees the cross as being the decisive moment in the battle between good and evil. Rather than the cross being a transaction between the aggrieved and the accused, it becomes the battleground between God and the devil, the place where the battle between good and evil reaches its pinnacle. Whereas in the "Latin" theory the cross was the sole responsibility of humanity and directed towards God, who was the recipient (object) of the crucifixion, for the "Classic" theory the cross is the sole responsibility of God and is directed towards the devil. Humanity is nearly completely (but, importantly, not absolutely) absolved of all responsibility for its deplorable situation, which is now seen as the deception of the devil. The cross is not a payment to assuage God's anger, but a siege to reclaim humanity from the clutches of the devil.

Christ, in a suspiciously docetic way, takes on the form of humanity and subjects himself to death. The devil, not being permitted to take Christ (who is God) in death, thereby "abusing his power" (Stott, 1986, p. 113), is thus defeated and the dominion he holds over humanity is rescinded. Christ's humanity, in this model of the cross, is intended to deceive the devil, and many theologians refer to the cross as a "trick that God had played on the devil, death, and sin" (Pelikan, 1977, p. 108). Or, as Gregory of Nyssa describes this same idea:

> In order that the exchange for us might be easily accepted by him who sought for it, the divine nature was concealed under the veil of our human nature so that, as with greedy fish, the hook of divinity might be swallowed along with the bait of flesh. (Gregory of Nyssa, 1977, p. 142; see also Stott, 1986, pp. 113–14)

In this way, the cross is not as in the first model—the paying of something *due* to the devil (which would legitimate his power to hold humanity hostage)—but is a "rescue mission". For the first model, God deserved the payment due to him as reparations, but in this model the devil is not due anything. The cross is God's way of winning back humanity by deception, not by paying anything due to the devil.

Criticisms of the cross as atonement

These views of the cross as atonement are not without criticism. This book has already made reference to the fact that evolution creates a problem for any theory of salvation, in that it strongly suggests that humanity (or any other creature) is not in need of salvation. Yet, these views of the cross have been criticized for other, theological, reasons.

a. Theological criticisms
Perhaps the most famous of these criticisms, and in many ways the most important, comes from Gregory of Nazianzus' "Oration 45: On Easter". Gregory asks:

> To whom was that Blood offered that was shed for us, and why was it shed? I mean the precious and famous Blood of our God and High priest and Sacrifice ... Now, since a ransom belongs only to him who holds in bondage, I ask to whom was this offered, and for what cause? If to the Evil One, fie upon the outrage! If the robber receives ransom, not only from God, but a ransom which consists of God Himself, and has such an illustrious payment for his tyranny, a payment for whose sake it would have been right for him to have left us alone altogether. But if to the Father, I ask first, how? For it was not by Him that we were being oppressed; and next, On what principle did the Blood of His Only begotten Son delight the Father, Who would not receive even Isaac, when he was being offered by his Father, but changed the sacrifice, putting a ram in the place of the human victim? Is it not evident that the

> Father accepts Him, but neither asked for Him nor demanded
> Him? (Gregory of Nazianzus, 1894, p. 431)

Essentially, Gregory is making two important points. The first is that the devil cannot be the object of the cross because this legitimates the power that the devil holds and would, therefore, seem to suggest dualism—God is not the only deity or divine power. As Young makes clear, "if we take God's primacy and uniqueness seriously, it would not do to suggest there is a powerful demonic being alongside God" (Young, 1992, p. 35). Not only does the postulation of a ransom to the devil suggest that God does not have sole or unique influence over God's creation, but it also compromises the "sort" of God that God is, i.e. not the eternal, non-spatial and non-temporal, simple God, but a God more akin to the Greek gods.

The second of these points, which has already been encountered, is that the Father cannot be the object of the cross because this creates a tension between the harsh justice of a God who demands recompense, and the love of a God who gives God's only Son to pay that recompense. A God who emptied and offered Godself is not the "sort" of God who would have demanded such a ransom or sacrifice to begin with. The God who so loved the world that God gave God's only Son does not describe a God who would demand such a sacrifice to satisfy divine justice in the first place. This does not even take into account the immutability of God, and the necessary change in God that the idea that the cross was a payment to the Father requires. If God is impassible and immutable, then nothing, not even the cross, can cause a change in God; therefore, God cannot *become* angry at creation and then subsequently be assuaged.

For Gregory, then, the question of the cross as salvation is not one of incompatibility with science (as it has become in recent years) but one of incompatibility with theology.

b. No salvation required

Another important criticism against theories of the cross as atonement is that modern science and, in particular, the neo-Darwinian theory of evolution, reject the idea that there is anything from which humanity (or, indeed, any creature) requires salvation. The most often cited reason for the need for the cross is that it offers a solution to the problem of sin. In this book we have already encountered the role that original sin plays in an evolutionary sensitive anthropology, and I have argued that what was considered to be negative from a theological point of view, is absolutely necessary from a biological one. Furthermore, Chapter 6, "Participation, Imitation, and neo-Darwinism", by noting that it understands the relationship between imitator and imitated in precisely the same way that neo-Darwinism does, supports this notion that the openness to failure and mistake is not to be considered a negative, but a neutral. If sin is understood as the reason for this openness to failure, then it cannot be anything from which humanity needs saving; on the contrary, it is precisely that which characterizes its ontology as creatures.

One of the consequences of sin, according to many theologians, is death. To be free from sin is to be free from death (Aulén, 1931, p. 67). Stott agrees with this connection, claiming that it is a central biblical motif. He writes that "throughout scripture, death is seen as a divine judgment on human disobedience" (Stott, 1986, p. 65). This means that, for the theologian, death is not a part of God's original intentions for creation; rather, it is a punishment on creation for its misbehaviour. McCord Adams attributes this idea to Augustine and Anselm, who both understood death as a punishment imposed for sin (McCord Adams, 1999, p. 14).

However, the neo-Darwinian synthesis (and the theory of evolution in general) disagrees strongly with this view of death as a temporary and unnatural phenomenon or as punishment. While Arthur Peacocke summarizes the view of many when he claims that pain and suffering "are present in biological evolution as a necessary condition for survival" (Peacocke, 2001a, pp. 31–2), this should not be taken as a reference to "survival of the fittest" or "red in tooth and claw". Rather, pain, suffering, and death are necessary in the sense that it is impossible to be created without being open to the failure and mistakes that inevitably lead to death.

Chapter 6, which discussed the role of participation and imitation as ontology and the connections that are present between it and neo-Darwinism, argued that the very fact of being created means imperfect replication. On the theological level this means an imperfect imitation of Christ, and on the biological level, it means genetic mutation. Neo-Darwinian evolution happens because that is precisely what it means to be created. Death is nothing but the logical conclusion of this openness to failure. Suffering and death are not things from which creatures require salvation, because it is impossible to be a creature and not be susceptible to suffering and death. Death is not a temporary and unnatural phenomenon imposed on creatures as punishment for their disobedience and poor behaviour; rather, it is nothing but the permanent, natural and inevitable manifestation of the very "imperfect replication" phenomenon that characterizes creation.

Other theories of crucifixion

These two theories of the cross are undoubtedly the most common and represent the vast majority of theologians and their understanding of what the cross means. However, there are other ways of understanding the cross.

a. Subjective atonement

The third of Aulén's categories (the Latin and the Classical were the first two), which he chiefly ascribes to Peter Abelard and Liberal Protestants such as Schleiermacher, is the "Subjective" model. As the name suggests, the difference between the "Subjective" model and the "Latin" and "Classical" is that for Abelard (and others who subscribe to this model) the cross does not achieve anything other than provoking a change of attitude in the individual. As Aulén describes it, the atonement consists of a "change taking place in men rather than a changed attitude on the part of God" (Aulén, 1931, p. 2).

The cross, for theologians who subscribe to this model, does not provoke an ontological or metaphysical change in creatures—such as removing the punishment of being susceptible to death—nor does it effect a change in the attitude of God to God's creatures. The cross, instead, is intended to produce a change in the attitude of the individual, whose new behaviour produces a new moral endeavour.

Despite the theological advantages of this model—i.e. it allows for the immutability of God that other models of atonement do not—there are obvious problems with it. If all that happens in the crucifixion is that creaturely attitude is changed, then this model is certainly open to the charge of Pelagianism. Of course, sin is no longer the context in which the crucifixion happens, but this still does not mean that creatures can be responsible for their own ontology either.

b. The cross as coronation

Thomas Schmidt, in his book *A Scandalous Beauty*, argues that the author of Mark's Gospel interprets the crucifixion as Christ's coronation. Drawing on many aspects of the coronation of Roman emperors, Schmidt shows how the author compares these aspects with the events of Christ's passion. Thus, for example, the purple cloak, the procession, and even the offering of wine, become Christ's very own coronation (Schmidt, 2002, pp. 31ff.).

Schmidt is not the only scholar to notice such a connection. Gregory Collins, for example, argues that Christ's exaltation begins "with his lifting up in crucifixion", so that the cross does not become Christ's execution, but his "throne" (Collins, 2010, p. 201). Hans-Ruedi Weber also sees the cross in such a light, finding this same idea in John's Gospel (albeit without the explicit imagery that Schmidt finds in Mark). Weber writes that John "interprets the crucifixion narrative wholly in the light of passages that proclaim his being raised up and glorified"; the execution is Christ's enthronement and the cross his throne (Weber, 1979, p. 132).

The intention behind such a comparison is easily understood. What is first thought to be God's ultimate failure, and the most obvious stumbling block to the mission of the Messiah, becomes precisely the opposite. The crucifixion becomes God's triumphant and glorious arrival to God's creation. In this way the crucifixion does not affect anything, as it does for other models, but is simply the moment that Christ's divine personality is made known.

c. Crucifixion and evolution

Another way of understanding the cross in the modern era is to see it as part of the process of evolution. Arguably the most important of these reinterpretations of the cross in this light is Teilhard de Chardin. Teilhard's theory of original sin has already been explored, and his theory of the cross follows from his theory of sin (as do most theories of the cross). Suffering, writes Teilhard, "is primarily the consequence of a *work in development* and the price that has to be paid for it" (Teilhard de Chardin, 1968a, p. 71). The cross, then, becomes God's payment for that progress: "the centre on which all earthly sufferings converge and in which they are all assuaged" (Teilhard de Chardin, 1968a, p. 67).

Put simply, "Christ is he who bears the burden, constructionally inevitable, of every sort of creation" (Teilhard de Chardin, 1969, p. 85). The cross is not God's way of removing that suffering, but of ensuring that the necessary suffering and sin in evolution are not allowed to overcome and smother the progress. The cross is God's assurance that the process of evolution will be successful and not be frustrated by the inevitable failure that is its consequence.

Jürgen Moltmann, another important twentieth-century theologian who attempts to deal with evolution, criticizes Teilhard's Christology on the basis that he has emphasized the creative aspects of Christ to the neglect of the redemptive. Disagreeing with Teilhard's emphasis, Moltmann labels his Christ, not the *Christus Evoluter*, but the *Christus Selector* (Moltmann, 1990, p. 294). If Christ controls evolution, so Moltmann argues, this means that Christ discriminates against those who are not "the fittest", and thus "[pays] no attention to evolution's victims" (Moltmann, 1990, p. 294). The cross then becomes almost the opposite of what it has traditionally been: it is the symbol of the strong and the successful and not the symbol of the weak and the downtrodden.

However, whilst it may be easy to be sympathetic to Moltmann's concerns for the *victims* of evolution, rather than affirming its *victors*, this theory leads Moltmann to claim that Christ is not just the "redeemer of humanity but evolution's redeemer as well" (Deane-Drummond, 2009, p. 46). For Moltmann, then, evolution is only a temporary phenomenon, and an undesirable one at that. In the end, Moltmann's theory is nothing more than a disguised theory of atonement; whereas for Teilhard, the process of evolution is positive—being used by God to create—for Moltmann it is negative, and something from which creatures need saving.

What is important about this model of the cross is that the cross becomes a theory of *creation* rather than *redemption*. This is crucial. The cross is about "reconstruction" and "re-creation" (Teilhard de Chardin, 1969, p. 145) rather than expiation; it "symbolizes much more the ascent of creation" (Teilhard de Chardin, 1969, p. 146). The atonement or salvation approach, therefore, becomes an interpretation of this "re-creation"; the cross as creation is more fundamental and primary, whilst the atonement becomes a lens through which this fundamental understanding is viewed (see also Murphy, 2013, pp. 71ff.).

Frances Young, disagreeing with Aulén's three models of atonement, also offers a theory of the cross that could be pigeonholed in the same category as Teilhard's. Drawing on the pericope in Romans in which Paul claims that "the whole creation has been groaning in labour pains" (Romans 8:22–3), Young sees the sufferings of the present age as the birth pangs of a new age (Young, 1992, p. 49). The cross, then, becomes "God actively bearing the brunt of the pain ... [thus] God suffers the birth pangs of his people, [and] takes responsibility for the appalling consequences of his act of creation" (Young, 1992, p. 57).

This idea of the cross as labour pains, as with Teilhard's theory of the cross as the catalyst for evolution, draws on the idea that creation requires suffering. Peacocke, for example, writes:

> Indeed the creation may be said to be through suffering, for suffering is widely recognized as having creative power when imbued with love. (Peacocke, 2001, pp. 87–8)

God cannot simply create—that creation must cost something. The cross is the price God pays for God's act of creation. This idea of the cross will influence the theory of the cross that this book will espouse.

The cross in a neo-Darwinian paradigm

It is clear, writes Palmer, that anyone who "thinks of [the cross] merely in terms of the remission of sins" does not understand it correctly (Palmer, 1949, p. 60). Not only is this understanding of the cross as remission of sins a limited and exclusive understanding, it is not even a fundamental one. Oliver Quick, for example, notes that it is possible to find "traces in Luke–Acts of a primitive Christianity which had not yet begun to understand Christ's death as a sacrifice for sin" (Quick, 1938, p. 220).

Gerald O'Collins notes the same in Mark, claiming that the "theme of atonement rarely surfaces in Mark" (O'Collins, 1983, p. 44). Rather, as it has already been suggested, the "notion of salvation through Christ [led] to the discernment of the plight from which rescue was needed" (Young, 2016, p. 8; see Williams, 2001, pp. 31–3 and pp. 80–1). Drawing on some of the criticisms of Augustine's theory of original sin, it was theologians who decided that humanity needed salvation and then read that atonement back into Christ's death.

The paradigm of atonement, then, is an interpretation of an earlier and more fundamental meaning of the cross. As it has already been suggested in relation to the evolutionary models of the cross, the cross is not about salvation and atonement, but about creation. Teilhard de Chardin points to this more fundamental element of the cross and writes:

> For obvious historical reasons, Christian thought and piety have hitherto given primary consideration in the dogma of redemption to the idea of expiatory reparation. Christ was regarded primarily as the lamb bearing the sins of the world, and the world primarily as fallen mass ... in addition, however, there was from the very beginning another element in the picture—a positive element, of reconstruction or re-creation. New heavens, a new earth: they were, even for an Augustine, the fruit and the price of the sacrifice of the cross. (Teilhard de Chardin, 1969, p. 145)

Of course, as has been a constant theme throughout this book, God cannot *re*-create anything because this idea implies both divine reaction to created events—i.e. that God is temporal—and multiple divine actions—i.e. that God is not simple. This means, importantly, that the cross is not a separate act of creation from the "initial" act of creation that happens through the self-emptying of the person of Christ in the incarnation. Indeed, there are many similarities between the incarnation and the crucifixion, and a comparison of the two can be mutually beneficial. The incarnation should be thought of as cruciform in nature and the cross should be understood incarnationally. The incarnation is the central event for all Christian theology, but it must be viewed through a cruciform lens.

The first thing that needs to be made clear when comparing the incarnation and the crucifixion is that, essentially, they are the same "type" of activity: both are "self-emptyings". Just as it was argued in Chapter 5 that creation and incarnation are the same "type" of event, so the incarnation and the crucifixion are the same "type" of event. Gordon Fee alludes to this connection through exegesis of the famous "kenotic hymn" in Philippians (Fee, 2002, pp. 80–1). Fee splits the hymn into two parts, the first describing the incarnation and the second describing the cross:

> [Christ Jesus] who, though he was in the form of God,
> did not regard equality with God as something to be exploited,
> but emptied himself,
> taking the form of a slave,
> being born in human likeness. (Philippians 2:6–7a)

> And being found in human form,
> he humbled himself
> and became obedient to the point of death—
> even death on a cross. (Philippians 2:7b–8)

What is important about this hymn is not just that it explicitly connects the two events, but that it uses identical language in the two parts: the language of the divine kenosis is mirrored in the divine crucifixion. One of the possible inferences of the kenotic hymn is that the two acts—incarnation and crucifixion—form two parts of the same divine act; the cross, in this interpretation, is not a *new* act of creation but is the continuation of the original.

Many theologians have noted this connection. Stott, for example, writes that the "self-humiliation" of the Son of God "began in the incarnation [and] culminated in his death" (Stott, 1986, p. 205), so that the "depths of condescension" are "not reached until the cross" (Shuster, 2002, p. 390). Likewise, Ferdand Prat writes that in the incarnation Christ "stripped himself" of divine honours, but "after the incarnation, the human will completes the self-stripping" (Prat, 1945, p. 319). The incarnation and the cross are connected by a shared content: both are concerned with the self-emptying or "humiliation" of Christ. This, as many theologians note, joins the events, seeing them as two aspects of the *same* self-emptying; or they understand the cross as being the end of a continual self-emptying that started in the descent from the Father's side, carried through his ministry, and ended in the descent into hell. In this way the cross "is kenotic self-emptying taken to its ultimate extent . . . it takes the incarnation to a new depth" (Deane-Drummond, 2009, p. 148).

What is significant about this connection is not the important nuances it makes to the doctrine of the cross (i.e. it shifts the focus away from atonement and towards creation), but the important confirmations it makes to the incarnation. The cross becomes the lens through which to view the incarnation, and this makes the incarnation a necessary "suffering" and "sacrificial" self-emptying "payment" for creation.

However, this does not mean that the crucifixion is that which gives meaning and purpose to the incarnation. The inference from the above quotations is that the crucifixion is the reason for the incarnation—that Christ only becomes incarnate as a means to the full self-emptying in the crucifixion. As Quick has already been quoted as noting about Pauline theology, the incarnation is a means of the atonement, rather than the atonement being a derivative of the incarnation, as with Johannine theology (Quick, 1938, p. 189). This Western, Pauline interpretation—with its characteristically blinkered preoccupation with legal rhetoric and salvation—has already been criticized. The incarnation, then, is not the beginning of a self-emptying that culminates in the crucifixion, which becomes the purpose of the whole process. It is the incarnation that is the starting point of theology, of which the crucifixion is a manifestation in time of the eternal divine act.

Furthermore, the incarnation cannot be the beginning of a process of self-emptying that culminates in the crucifixion, because this places unacceptable temporal categories on God. The simplicity of divine activity, which has been an important element of how neo-Darwinism influences Christian doctrine, demands not only that salvation be rejected (so there is not creation *and* salvation, but only creation), but also the idea that the divine action can perdure in time (i.e. start a process of creation in the incarnation for completion in the crucifixion in the future). The singularity of divine activity is not a uniting of distinct acts through intention, but a singularity of event—i.e. only one divine act, which God does once. This means that the cross is not the end or completion of creation—as if the incarnation and crucifixion form two ends to one process—but they must be the *same* event.

———

Claiming that the incarnation and the crucifixion are the same event does not deny that the crucifixion definitely does happen after the birth of Christ—thirty-three years later, in fact.[32] This is not a claim for the eternal nature of a definitely temporal event. The incarnation (i.e. the self-emptying of the divine nature) is an eternal event; it has a temporal anchor (first-century Palestine), but temporal language cannot be applied to it—there is not a time when the Son is not self-emptying, either before or after the incarnation.

However, the crucifixion (i.e. the self-emptying of the created nature) has a definite temporal locus and can be described by temporal language; there is a definite before and after of the crucifixion. In this way, the crucifixion, despite the singularity and simplicity of divine activity, does "continue" the incarnation and does "carry on" self-emptying after the incarnation. The point, however, is that this does not imply that the incarnation is somehow incomplete without the crucifixion, as if the crucifixion finishes what was started in the incarnation. Rather, the crucifixion is the temporal pole or locus of the incarnation: it is the temporal created activity that mirrors and reveals the eternal divine activity. Gregory Collins perfectly illustrates this approach, when he writes:

> The kenosis of Christ on Calvary is a window opened out by God, revealing the kenosis going on eternally "within" the life of the Holy Trinity. Christ's earthly cross points us beyond itself towards the "cross" of undying love set up eternally within the heart of God. (Collins, 2010, p. 285)

It is important to criticize Collins here, in that it is only the Son who self-empties, not the Father and the Spirit, although they are included in the single divine act. However, Collins is correct that the cross, because it is an action of the God-creature, must give an insight into what happens in the incarnation. The cross does not complete the incarnation—the cross points to the incarnation and interprets it; yet, in doing so, it is identical with that event as well.

a. *Communicatio idiomatum*

The doctrine of *communicatio idiomatum* can help to nuance how it is that the incarnation and the cross can be seen as the same event, rather than one being a continuation of the other. Jesus' death can still be described as the consequence of the eternal divine activity without the problems of temporality, or how it is that the eternal can experience loss, because of the relationship between creation and divinity in Christ's person. Likewise, the cross can be seen as being identified with the eternal incarnation through *communicatio idiomatum*. This doctrine, then, also helps to solve the question of how an act in time can be an eternal divine action. It helps to answer how it is that the incarnation and the cross are one and the same action.

The incarnation is, properly, the work of the divine nature and the cross the work of the created; the former is eternal and cannot be applied to creation and the latter is temporal and cannot be applied to divinity. However, the doctrine of *communicatio idiomatum* "conflates" the two; both are the action of the *one* subject. Both are, at their simplest, the result of kenosis, and as a result are easily shown to be applicable to each other, and thus to be identical.

If *communicatio idiomatum* describes how the creature creates and the divine dies, then this can be reworded to mean that the *communicatio idiomatum* describes how the divine crucifies and the creature incarnates. Both creating and dying are the same act; the divine kenosis takes the form of incarnation and the created kenosis takes the form of crucifixion. These are not two acts, but the same act—the same, single, simple act of the one person; from the divine side it looks like incarnation and from the created side it looks like crucifixion. Two energies, two manifestations of two wills, but completely and perfectly united (by the complete and utter submission of the created to the divine) in one subject—nowhere is this better illustrated than in the Garden of Gethsemane.

b. Sacraments

The sacraments, or more precisely the relationship between the Mass and Calvary, can also demonstrate the relationship between the eternal self-emptying of the incarnation and the temporal self-emptying of crucifixion. Risto Saarinen writes about the Mass as follows:

> The unique sacrifice of Christ is not repeated in the Eucharist . . . the Eucharist does not bring about anything in addition to the once-and-for-all atonement. (Saarinen, 1997, p. 27)

Teilhard de Chardin also claims that "all the communions of our life are, in fact, only successive instants or episodes in one single communion" (Teilhard de Chardin, 1969, p. 166) and that "all the communions of all men now living are one communion. All the communions of all men, present, past and future, are one communion" (Teilhard de Chardin, 1968, p. 124). This means that the Mass is not a new or separate crucifixion; all Masses are just different temporal manifestations of exactly the same event, identical with the sacrifice of Calvary. In this way, the crucifixion, temporally distinct from the incarnation, is still identical to it. The crucifixion, it might be claimed, is a sacrament of the incarnation.

In this way, it is possible to see the sacraments, not as "re-presentations"[33] of the crucifixion, but as "re-presentations" of the *incarnation* (or, at the very least, as "re-presentations" of the crucifixion, which in turn is a "re-presentation" of the incarnation). This connection of the sacraments with the incarnation is at times explicit. P. Sherrard, for example, writes that "each sacramental action and object is the incarnation" (Sherrard, 1964, p. 137). N. Zernov, too, writes that "the entire story of the incarnation is enacted during each Eucharist" (Zernov, 1964, p. 118).

Not only do these quotations add further support to the idea noted above, that each Mass is not a repetition of the self-emptying of Calvary but is identical with that event, but they also further support the idea that there is a fundamental connection between what happened on Calvary and what happens in the incarnation. Indeed, the epiclesis, by being the moment that the bread and the wine become the body and blood of Christ, is a sort of incarnation.[34]

There is not the room to discuss the full sacramental implications of these connections. The purpose of mentioning them is to further illustrate the identity between the self-emptying of the incarnation and the self-emptying of the crucifixion and, more importantly, how they can be numerically identical while temporally distinct. The incarnation is the eternal divine activity of the divine nature and the crucifixion the temporal created activity of the created nature: distinct but perfectly united. The relationship of the cross to the incarnation is exactly the same as the relationship between the Eucharist and the cross.

The cross, then, is not a distinct act—a *re*-creation—that atones for, and resets, the sin of humanity; the cross is the temporal pole of the incarnation, the illustration of the perfect union between the divine and created natures/wills/energies in Christ. The cross is the acting out in time of the eternal divine act of creation.

The solidarity of Christ with creation

One of the important elements of the theology of the cross is the idea that the crucifixion is God coming down to experience the pain and suffering of God's creatures. This "suffering in solidarity", however, is not just a suffering with creatures; but by suffering with creatures, this suffering becomes transcended and infused with ultimate meaning. In this way, Young writes:

> The cross is now widely treated as a sign of God so loving the world that in Jesus Christ God's very self identified with, indeed entered into, that surd of suffering that seems to challenge God's goodness and the goodness of the divine creation, God thus taking responsibility for it all and suffering alongside . . . it is as if from the beginning the creator of all bore the consequences of a creation in which there was potential for death as well as life. Thus, the cross cut across time and space, finding its location in God's eternal purposes, no mere remedy for an unfortunate accident within time, but a feature built into the very structure of the universe and a sign that "the creator intended to use death creatively". (Young, 2016, pp. 98–9)

Of course, as in the neo-Darwinian-influenced theology of divine activity outlined in the previous chapters, God cannot "use death creatively" any more than God can use evolution creatively; the whole paradigm of divine influence/providence of creation is rejected. However, the point of the quotation is to demonstrate the centrality of the idea that God in Christ shares in the suffering of God's creatures. This, however, is at odds with the idea of the incarnation as *primary* and as giving birth to the *firstborn* of creation. If, as it has already been noted, Christ is the firstborn of all creatures, whose incarnation is the act of creation that is ultimately responsible for giving being to every creature, then Christ cannot share in the life of the creation. Rather, it is the complete opposite: it is the sharing in the life of Christ (i.e. participation/imitation) that is the basis of a Christian ontology. Christ, in the incarnation, does not assume something that is *already* existent; the incarnation creates an ontological space, in which all creatures participate/imitate, and is thus the basis of their existence. The incarnation is not a sharing in the human condition; being created is a sharing in the "Jesus condition".

This means, therefore, that the suffering that Christ undergoes through his crucifixion is not a sharing in the suffering of creatures; rather, *the suffering of all creatures is a sharing in the suffering of Christ*. Just as the incarnation is primary and the ground of all being, so the crucifixion must be primary and the ground of all being. Just as the crucifixion (as a creating self-emptying) points back and provides the lens through which the *incarnation* must be viewed, so the crucifixion provides the lens through which the *ground of being* must be viewed. Put simply, then, the crucifixion is the pattern according to which all creatures are created: to be created is to be open to the failure and mistake that the universal ontology of participation/imitation (of which neo-Darwinism is a biological manifestation) describes. To be created, in other words, is to suffer.

Once again, this is not a novel approach to theology and many theologians have noticed this element. Teilhard writes that "the universe assumes the form of Christ"—which this book argued for extensively in the previous chapter—"but, O mystery! The man we see is Christ crucified" (Teilhard de Chardin, 1968a, p. 208). Michael Gorman, too, claims that "the cross is what creation is all about" (Gorman, 2009, p. 34), as does Alister McGrath, who claims both that the cross is the "source of ideas about God" and "the natural pattern of Christian life" (McGrath, 1987, p. 165), suggesting a theology of the cross very much like the one outlined here.

The cross is both the lens that illuminates the invisible incarnation—and thus reveals the single and simple divine activity—and the lens that illuminates what it means to be created—and thus reveals the ontology of imitation/participation of which neo-Darwinism is a biological manifestation. The creature is called to imitate suffering because the creature already imitates suffering as the very ground of its being—indeed, suffering is the very explanation of imitation in the first place (i.e. imitation is imperfect copying, which it was argued earlier, is identical to the openness to failure).

Just as the incarnation was argued to be responsible for participation, so the crucifixion—still a self-emptying but more explicitly or obviously a suffering self-emptying—demonstrates the openness to failure that characterizes that participation. Causing creatures to participate in God, then, is the same event as the incarnation/crucifixion, just seen from different perspectives; participation is the effect as the incarnation/crucifixion is the cause—"the divine calling and the human responding are virtually the same activity viewed from different perspectives" (McIntosh, 1998, p. 221).

The cross, then, becomes that which the follower of Christ is called to imitate. Creatures already suffer as the very basis of their existence, and they are called to suffer as the way to follow Christ. The centrality of martyrdom (including the "white martyrdom" of monasticism [Williams, 1979, p. 92]) and asceticism give credence to this centrality of suffering as the image of Christ that all creatures imitate, and nothing more is needed here to illustrate just how crucial it has been to Christian theology. What is important to note is that the conclusion that to exist is to suffer, that it is impossible to be created and not suffer, and that creatures are called to embrace this suffering as something to strive towards, can be—and should be—a positive conclusion.

Suffering is not just the way to find God—suffering *is* God. Suffering is not a means to an end, it *is* the end of Christian life—in the same way that imitation is not a means to an end but the end in itself. Just as was shown in the previous chapter, the subjective nature of imitation means that creation and deification are precisely the same from the objective (i.e. divine) perspective, but entirely different from the subjective (i.e. created) perspective. That is, it is only the subject that can differentiate between creation and deification, multiple events for the creature and single event for God—so suffering can be seen by one creature as abandonment and distance from God and another as deification, as proximity to God. Thus, Collins writes:

> The crucial point is this: the darkness is *itself* the irradiating ray that guides one forward for it is God's own presence. It is grace. That's why it is folly to want to escape the darkness, leave it behind or imagine that having "once" accepted it and "got over it", one will then be admitted to the titillating rewards of a new sensory or intellectual "illumination" . . . learning that the darkness itself is light means entering deeply into the mystery of Christ and sharing in his transforming passage through death. (Collins, 2010, p. 241)

CHAPTER 8

The Resurrection

The previous chapter, ending as it does by claiming that suffering is not only an insurmountable condition of being created but also the very basis and grounding of existence, reaches a very depressing conclusion. Suffering is what it means to be created; suffering is the inevitable result of being open to failure and mistake, which is the very basis of being created, both from a biological (i.e. neo-Darwinist) and a theological (i.e. participation/imitation) point of view. However, in as much as Christology is incomplete without a discussion of the cross, so a discussion of the cross is always incomplete without reference to the resurrection; the cross is not the end of the story. As Alister McGrath writes:

> Without the resurrection, the way of the cross is nothing more than ascetic self-denial, at best a way of resignation to the futility of existence, at worst a way of despair and delusion. (McGrath, 1987, p. 165)

The cross without the resurrection can never be the final word on God and Christ's interaction with creatures.

Yet, while it is often assumed that the theory of evolution provides the biggest challenge for theology by questioning the doctrine of creation, it is actually the doctrine of the resurrection that provides a bigger challenge. Many theologians (whether rightly or wrongly), including the theology presented in this book, have provided possible ways in which the theory of evolution, and in particular the neo-Darwinian synthesis, can be seen as what *should be expected* from a theory of creation. The doctrine of the resurrection, on the other hand, is a stumbling block to a full acceptance of the neo-Darwinian synthesis.

Neo-Darwinism disagrees with the resurrection on two very important, but related, points: firstly, it rejects the possibility of being created *without* being open to failure and mistake and secondly, it rejects the inherent teleological nature of resurrection. On the one hand, the doctrine of resurrection from the dead, understood as the future state of creation in which suffering and failure play no part, is directly opposed to the permanent openness to failure that is characteristic of the neo-Darwinian synthesis. On the other hand, the future aspect of this "perfect" state of creation is opposed to the lack of teleology in neo-Darwinism. Evolution, claims the neo-Darwinist, does not work towards a goal; change happens accidentally and is replicated imperfectly.

The question that clearly presents itself is to what degree is the resurrection meaningful in a neo-Darwinian paradigm and what does that resurrection look like when the possibility of an immutable and future body is rejected as incompatible with an ontology on which both biology and theology agree?

This chapter will answer that question by considering a number of important elements regarding the doctrine of the resurrection. It will show how, with only a few select nuances and, like other chapters, drawing on and re-emphasizing elements that are *already* present in Patristic and Scholastic theology, a doctrine of the resurrection is possible in a neo-Darwinian paradigm that seems to reject the very core elements of the doctrine.

—

The first section of this chapter will begin by showing that the popular understanding of the resurrection must be rejected on biological, physical, and philosophical grounds. The permanent nature of imperfect replication, the non-teleological nature of evolution, the role of death in a four-dimensional universe, and the problem of personal identity all contribute to the rejection of the resurrection as the future raising from the dead of a perfect body. This is followed by a consideration of the resurrection in the New Testament, which will argue that there are similarities between the Genesis narrative and the resurrection accounts in the Bible, in the sense that both contain multiple and contradicting versions. This means that, as with Genesis, an analogical and metaphorical interpretation of the resurrection is possible.

The second section will describe a different and nuanced interpretation of the resurrection. This will allow for a more subjective approach to the doctrine of the resurrection, one in which the salient and crucial factor is not the preservation of life, but relationship with God. This, it will be concluded, sets the discussion of the resurrection in the same language as that of the discussion of participation and imitation.

Interestingly, therefore, this chapter could be understood as a re-exploration of the same themes that were discussed in Chapter 6, "Participation, Imitation, and Neo-Darwinism". That chapter was about the creature's relationship with God as the ground of its being, and this chapter makes essentially the same point. As Young points out, "birth and resurrection are intimately related ideas"; there is a parallel between creation and resurrection (Young, 1992, p. 107). In a sense, therefore, there is a similar relationship between the cross and resurrection, as there is between incarnation and participation. The cross, which further explores the role of the incarnation, leads to the resurrection, which further explores the role of participation.

Problems with resurrection

The resurrection of the body from the dead is a central Christian doctrine. It claims that the body of the deceased will be raised from the dead incorruptible, into a new life that is free from suffering and, the problem of sin being overcome in Christ, closer to God. However, advancements in scientific knowledge, both in biology and physics, have made belief in this doctrine very problematic. From both a biological and a physical point of view the idea of a future and incorruptible body is highly questionable.

However, the idea of resurrection as the future raising from the dead of a perfect body is not just questionable on scientific grounds. Those scientific problems provoke theological questions that support the scientific questioning of the resurrection. Furthermore, there are also philosophical problems with the idea that the personal identity of the creature can be continued after death. It is not just that science questions the future incorruptible body; philosophy also questions whether that future incorruptible body could solve the problem for which it was proposed to solve, i.e. the ability to survive death.

a. Resurrection and neo-Darwinism

As it was claimed earlier, whilst it is popular to see creation as the biggest problem posed by evolution, it is actually resurrection with which evolution has the biggest problem. The reason for this, as I argued in Chapter 2, "Theological Anthropology", is that the theological description of what it means to be created—i.e. original sin and openness to failure—is mirrored in the "imperfect replication" of neo-Darwinism's description of what it means to be created, thus evidencing a certain agreement between theological and biological ideas of creation. However, that "imperfect replication"—which accounts for the openness to failure that characterizes what it means to be created—is a *permanent* feature of creation. The permanence of this openness to failure means that the perfection or infallibility that characterizes the resurrected body is not allowed by neo-Darwinism. There is simply no way to be created, according to neo-Darwinism, without being open to failure and mistake; or in other words, without being mutable, passible, and corruptible.

Stephen Davis describes the resurrected body as "immune to evil", unable to suffer, "will not grow old or die", "will need no material food or drink", and as possessing superpowers such as "agility"—which is the power to "come and go at will, unimpeded by things like walls and doors" (Davis, 1993, p. 94)—and in doing so describes a popular conception of what the resurrected body is like. It is clear that these characteristics are completely rejected by the permanent corruptibility of neo-Darwinism. As Ray Hart writes:

> What does the human person most deeply yearn for, long for? One might answer "infallibility", but that answer is useless because whatever infallible being might be it would not be human being, in whom freedom means minimally the latitude of possible failure. (Hart, 1997, p. 134)

To be infallible is to be not created.

—

Resurrection as a future raising from the dead of a perfect body is also denied by neo-Darwinism on the basis that it rejects the importance of "the future" as a category. One of the salient features of Darwin's theory of evolution, which was why he was dissatisfied with his theory being called "evolution" (Birx, 1991, p. 133), was that he strongly disagreed with the inherent progress that characterized all other theories of evolution. For Darwin, evolutionary change is not a temporary process that is implemented in order to bring about perfection, which will subsequently stop once that perfection has been reached. As I have already extensively argued, evolutionary change is simply the observation that genetic replication is fallible and that the change that is produced allows for a differential survival and reproduction rate. In other words, evolution, according to neo-Darwinism, is teleologically neutral; evolution has no goal towards which it is heading.

This means that resurrection as a *future* reality is denied because the future is no more perfect than the present. The future is different, but not closer to being perfect. In this way, the future is not closer to God. Death, to which resurrection is postulated as a solution, no longer excludes the individual from a state of perfection because that state of perfection is impossible, both as a present and a future reality.

b. Resurrection and physics

The resurrection as a *future* raising of a perfect body from the dead can be rejected from a physical point of view as well as a biological one. The future is not just rejected as a useful category by the biologist; the physicist also has problems with the idea that the future is in some way more real or closer to perfection. In this way, the physicist disagrees that a future raising of the body from the dead can bring the individual creature closer to God.

The neo-Darwinian synthesis, it has already been noted, completely rejects the idea—so common in theological accounts of evolution—that the universe is moving or tending towards a perfect creation, or even simply a better creation. This means that resurrection as a future raising from the dead is not needed, because death does not exclude the individual from closeness to God on the basis that there is not a future time that is closer to God. However, the physicist adds to this by claiming that not only can the future never be considered better than the present, and thus death does not exclude the individual from greater relationship with God or greater perfection, but death does not even remove the individual creature from existence.

For the physicist time is simply an "illusion" (Rovelli, 2015, p. 58), which means that the past and the future are not inexistent, but are simply inaccessible to the creature in the same way that different spatial locations are inaccessible to the individual. This conclusion is reached due to the fact that time is simply another dimension, in the same way that space is. The creature does not occupy a three-dimensional universe that perdures through time, but occupies a four-dimensional universe in which it moves through time in a way that is strikingly similar (although not identical) to the way the creature moves through space. All times are equally "real" and "existent", in the same way that all places are equally "real" and "existent".

This means that the non-presence of an individual at a particular time presents only the same problem as the non-presence of an individual from a particular place; two individuals who are separated by space (for example in London and Jerusalem) are separated in much the same way as two individuals who are separated by time (for example, first century and twenty-first century). Death, then, becomes only a boundary to existence, in the same way that the individual has spatial boundaries. Time is not a constant moving in and out of existence of all matter—it is a boundary that gives meaning to matter in precisely the same way that space does.

The narrator in the novel *Slaughterhouse Five* describes the role of death in this four-dimensional universe: "when a person dies, he only *appears* to die. He is still very much alive in the past" because "all moments, past, present, and future have existed, [and] always will exist ... it is just an illusion we have here on earth that one moment follows another one, like beads on a string, and that once a moment is gone it is gone forever" (Vonnegut, 2000, p. 22). Thus, a person, "according to the Earthling concept of time" who had died back in 1958 "was still alive somewhere and always would be", according to the "Tralfamadorian" idea of time (Vonnegut, 2000, p. 164).

This description may be from a work of science fiction, but it describes a very real physical idea. If modern, Einsteinian physics is correct, then the "Tralfamadorian" concept of time is an accurate description of the universe. It is only because of the individual creature's (or "Earthling's") experience of time that this nature of time goes undetected; just because creatures experience time in a different way to the way they experience space, does not mean that time is a different "thing" to space.

The idea of resurrection as a future reality is rejected, therefore, because, not only is the future no longer "more" existent or greater than the present, but death does not remove the individual from existence. The creature that has died is not removed from existence; the creature still exists but in another inaccessible temporal location—in precisely the same way that the creature in Jerusalem exists in another inaccessible spatial location. Resurrection, then, simply cannot be construed along temporal grounds, because the past is not inexistent and the creature that has "died" is still alive somewhere and "sometime"; the creature does not need to rise from the dead because the creature is not dead!

c. Resurrection and theology

However, not only does the idea of resurrection as a future reality contradict physical principles, it also contradicts theological principles. If the resurrection is a future reality, and that reality is characterized by a closer relationship with God, then the suggestion that resurrection is a future phenomenon would have to include the claim that God is more present to, or closer to, the future. This would have to mean that God is temporal. Claiming that the resurrected body is closer to God is precisely the same as saying that God is closer to the future, and, thus, claiming that God is susceptible to physical dimensions.

In other words, even if the biological and physical criticisms were incorrect, and there were a future resurrection, this would still be disagreeable on *theological* grounds, because this would make God locatable in time; God would be there and not here (or at the very least, more there than here). Due to the mutual inclusivity of time and space, claiming that God is closer to the future would also require God to be closer to a particular space. Quite simply, then, the idea of resurrection as a future phenomenon that places the individual in a closer relationship with God makes God a creature. The resurrection, understood as a future perfect body, does nothing but make God susceptible to physical dimensions.

Thus, the universe does not move or tend towards perfection. Further, death is simply a temporal boundary; it does not remove the creature from existence, let alone exclude it from a future perfection or closeness to God. From a purely physical point of view death is not a problem that requires a solution, as it no longer excludes the creature from anything. However, even if there were a future raising from death to perfection, this could not bring the creature closer to God, because God is not temporal. Therefore, it is not so much that the future raising from the dead is actually biologically or physically possible—i.e. that a future and incorruptible body is possible—as to deny that it achieves what it is supposed to—i.e. bring the creature closer to God.

d. Resurrection and personal identity

Another problem with the idea of the resurrection as a future raising from the dead is the question of the continuation of personal identity. Put simply, if the body that is raised from the dead cannot be identified as the same person who died, then the resurrection has not occurred. There needs to be a single personal subject that describes both the corrupted and dead body and the incorruptible and resurrected body, or the resurrected body is useless to the dead person and offers no solace. As Joseph Ratzinger asks, "How can there be any identity between the human being who existed in the past and the counterpart that has to be recreated from nothing?" (Ratzinger, 1988, p. 106).

This is nothing but a re-presentation of the famous "ship of Theseus" problem—a philosophical conundrum that has been around since Plutarch posed it more than two thousand years ago. The crux of the problem is whether a ship that has had every part replaced can truly be the same ship. Its relevance for the problem of resurrection is obvious: if the body undergoes a complete decay, and a new body is raised at the resurrection, in what way is it the same person? It is noteworthy that the problem exists for both transformation and replacement theories of resurrection. While there is not the room here to delve into the intricacies of the problem, there are essentially two possible solutions. Either the raising from the dead is a literal re-gathering of atoms, or personal identity is attributed to the presence of "something else", normally a soul.

The first way of dealing with the problem is to claim that the atoms from which the body is made are reconstituted; however, this has major difficulties. In this direction, Hans Küng writes that there is "no continuity of the body" (Küng, 1978, p. 251). Modern scientific advancement has discovered that the atomic and chemical make-up of the body is never static, as the body goes through many different chemical changes over a lifetime: blood is renewed, skin is shed, and so on. This means that the individual, personal subject does not possess only one body, but many different atomically and chemically distinct bodies over the course of a lifetime. The question of the resurrection of the body, then, provokes a very serious question: which body is resurrected?

There is, furthermore, the fact that organic material is used to keep other creatures alive. Not just predation, but plant matter as well; all organic matter is used as nutrition for other organic material. This means that it is not just that there is a massive problem over what particular manifestation of atoms and chemicals constitutes a particular creature, but all of those atoms and chemicals will have constituted many other particular creatures in the past and will do in the future. The future raising of the body from the dead is rejected, then, on the basis that there is simply no way to determinate what atoms and chemicals are pulled together and whether those atoms and chemicals are not shared by other creatures.

The second way of dealing with the problem of the continuation of personal identity in the resurrected body, which can also deal with the problems of continuing flux of the atomic and chemical make-up of the body, is to postulate that "something else" maintains personal identity. Normally, this is called the soul, i.e. a non-material "something" that gives life to the disparate elements of the body and maintains and secures personal identity. The immaterial soul survives the death of the body and reanimates another body (however similar it is to the first; the apostles had trouble recognizing Jesus to begin with), or a transformed reanimation of the first (the empty tomb in the gospel narrative suggests that Jesus' resurrection at least was a reanimation of the same body).

However, this has already been completely rejected. Cosmological dualism is denied by the neo-Darwinian synthesis and this must include anthropological dualism as well. There is simply no room in any theology that wants to take seriously the neo-Darwinian synthesis for the appeal to a soul or anything immaterial.

Therefore, resurrection as the future raising of the body to perfection is denied on the basis that there can be no way to maintain and secure the personal identity of the individual subject between the death of the creature and the future raised body. The subject does not possess one distinct body, not to mention that those distinct chemicals and atoms constitute the bodies of many different creatures. Secondly, the personal identity of the individual cannot be reduced to an immaterial soul. The neo-Darwinian synthesis strongly denies the presence of the immaterial soul and must likewise be seen as a rejection of the resurrection as a future incorruptible body.

The resurrection and the New Testament

The resurrection as a future raising of an incorruptible body is rejected from biological, physical, and philosophical positions. It has also been shown that these rejections are in keeping with the emphasis on the non-teleological nature of God that I have maintained throughout this book; if God is not temporal, then proximity to God cannot be expounded in terms of a future reality. However, the emphasis on the eternity of God and its relationship to creatures is not the only theological reason why such an understanding of the resurrection could not be supported; the Bible also questions the "classical" or "traditional" idea of resurrection.

The New Testament is crucial as evidence for the occurrence of the resurrection of Jesus Christ, and it provides the locus around which all discussion of the resurrection orbits. However, as with most questions concerning the Bible in the modern world, the relevant issue here is one of interpretation: in what sense should the resurrection be taken? Are theologians compelled to take the resurrection narratives at face value, as purely historical accounts of what actually happened at the time, or is the resurrection a theological motif that describes a theological point? Hans Küng seems to answer this question, alluding to the non-historicity of the resurrection narratives, writing:

> The New Testament authors are not interested in any kind of completeness nor in a definite sequence and least of all in a critical historical investigation of the different pieces of information. From this it is clear that there is something more important to be stressed in the individual narratives. (Küng, 1978, p. 348)

Küng argues that the New Testament authors, when discussing the resurrection, are not seeking to give historically accurate accounts of what transpired in first-century Palestine; rather, they are using allegorical language to make a theological point about the meaning and significance of Jesus. The point of the resurrection stories is not to describe what happened to Jesus, but to be a "divine seal of approval" of everything that Jesus is and did (Lane, 1975, p. 71). For Küng, then, criticizing the Gospels for presenting historically inaccurate events misses the point; the question of historicity is not important to the New Testament, whose primary intention is to make a theological point. Therefore:

> All questions about the historicity of the empty tomb and the Easter experience cease to count beside the question of the significance of the resurrection message. (Küng, 1978, p. 379)

The resurrection is about creaturely relationship with God, not the future raising from the dead.

The insignificance of the historical accuracy of the resurrection story is easily seen in the contradictions and inconsistencies of the different narratives. Thus, many commentators on the resurrection story in the New Testament note that there is a difference in the presentation of the narrative in the Gospels (Barclay, 1996, p. 19)—disagreements over the visitors to the tomb, who was present at the tomb, and what happened at the tomb (see Brown, 1973, pp. 99ff. and p. 118). The question of exactly what happened is simply not an issue for the New Testament authors and so different elements of the story are nuanced in order to emphasize a particular theological or Christological point.

The same phenomenon has already been noted regarding evolutionary commentary on the Genesis narrative. Even without the criticism over the quality of the science recorded in Genesis, the contradictions contained within the opening chapters are enough to convince the reader that what the editor (assuming that there is not a single author of Genesis) is attempting to do is not give a historically accurate account of the creation of the universe, but present a very important theological point, i.e. all creatures depend entirely and solely on God for their existence. Likewise, the same conclusion was reached in relation to the doctrine of original sin. The mythical language of the Genesis narrative was rejected, but the point of the myth was upheld by the neo Darwinian synthesis. In other words, denying the historical accuracy of the Genesis narrative does not deny the theological truth of what the narrative is trying to portray.

These same conclusions can be applied to the resurrection accounts. The contradictions in the resurrection stories, which are so obvious that they could not have escaped the notice of the compilers of the New Testament, only serve to emphasize that the point of the stories is not to present a historically accurate account, but to make a theological point: that the purpose of human life is relationship with God. The New Testament is important, therefore, not as historical evidence but as theological interpretation.

What is resurrection?

With the idea of a future raising of an incorruptible body rejected by biology, physics, and philosophy, and not even demanded by biblical theology (which suggests a more allegorical or metaphorical understanding of the resurrection, as with the Genesis narrative), a different and nuanced understanding of the resurrection is required. As it has already been claimed, the important point about the resurrection is that it facilitates a closer relationship with God. Oliver Quick, in his book *Doctrines of the Creed*, makes this clear, writing that

> [t]he object of ultimate hope is communion with the eternal God, and not any prolongation of human lives as such. (Quick, 1938, p. 262)

The resurrection, for Quick, is not primarily about the raising of the body, but what the raising of the body means for the creature's relationship with God. The content of the doctrine of the resurrection, then, is about relationship with the divine, not human perfection.

For the "classical" approach to resurrection, that closer relationship with God *depends* upon an anthropological condition. Or, to put this point differently: if the future raising of a perfect body does *not* provide a closer relationship to God then the raising is pointless; as Georges Florovsky makes clear, only communion with God and life in Christ makes "the restoration of human wholeness gain meaning" (Florovsky, 1976, p. 151). The reason that closer relationship with God was bound up with bodily perfection was that, at least for Western theologians such as Augustine, death was introduced into creaturely existence as a result of (or punishment for) the marring of human relationship with God. When this understanding of sin and death is reinterpreted in the light of neo-Darwinism, this interpretation of the resurrection can also be nuanced. The resurrection is about communion with the divine, *not* the prolongation of creaturely existence. The prolongation of creaturely existence is only a means to the end of communion with the divine, which is the real point of resurrection. Or, as Ray Anderson puts it, the resurrection "has a soteriological rather than an anthropological significance" (Anderson, 1986, p. 44). Closeness to God is no longer bound up with bodily perfection or incorruptibility because death is no longer related to relationship with God as the consequence of sin.

In emphasizing the resurrection as proximity to God, it is argued that the accompanying element of a future and incorruptible body is superfluous. The resurrection is a theological doctrine, not a biological one. Just as evolution does not change the central element of creation (i.e. creatures are entirely dependent on God for their being), but simply shifts the paradigm and reframes the language, so neo-Darwinism does not change the central element of resurrection (i.e. that through Christ, creation can come to a closer relationship with God) but simply shifts and reframes that central core.

Resurrection as subjective experience

The resurrection, then, is not an objective future raising from the dead. Rather, it is a subjective interpretation of the significance of Christ and the realization that the individual already imitates Christ as the very ground of their being. In this way, the resurrection is not something that happens to somebody *after* death, but *during* "this life". Willi Marxsen, for example, writes:

> Jesus lived and gave a resurrection into new life even before his crucifixion. One could even say that Jesus was risen before he was crucified ... [John] uses the word ["life"] to express both the earthly life and the heavenly life, whereby it is noteworthy that eternal life is not the after-life but a certain kind of life *in* this earthly life [cf. John 5:24]. Another feature is peculiar to this Gospel. The traditional concept of the expected *parousia* is shifted to the present ... the *parousia* takes place wherever Jesus' word is obeyed ... here [John 17:3] the future aspect of eternal life threatens to disappear entirely. (Marxsen, 1970, pp. 184–5)

The resurrection is not a bodily transformation but a mental transformation; the resurrection is coming to see the world differently. The resurrection is not an objective transformation of the body after death, but a subjective transformation of the perception and meaning of "this life" before death. Not only is the resurrection shifted away from a future reality into a present one, but the explicit reference to obeying Jesus' word means that it is possible to understand this present, subjective resurrection in terms of imitation of Christ; the resurrected person is the person who imitates Christ.

a. The Transfiguration

The Transfiguration of Christ can help to clarify this notion of the resurrection as subjective experience. Many theologians note the connection between the transfiguration and the resurrection, claiming that the "power" (Marshall, 1994, p. 20) or the "glory" (Bonaventure, 1978, p. 135) of the resurrected Jesus was momentarily revealed on Mount Tabor before being permanently revealed in the resurrection. In fact, David Brown goes even further by claiming that "the Transfiguration is in fact a transposed original resurrection appearance" (Brown, 1985, p. 192). The identification of the transfiguration and the resurrection means that what is claimed about the transfiguration can contribute to the doctrine of the resurrection.

In this direction, then, what is important about the transfiguration is not so much that it removes the connection between Christ's death and his resurrection—such that the resurrection occurs before Christ's death and is no longer a remedy or solution to the problem of death (i.e. concerned with prolongation of creaturely life)—but that it also removes the objective element of the resurrection. In fact, not only is the transfiguration/resurrection not an objective transformation of Christ, it is not even something that happens to Christ. No change takes place in Christ when his glory is revealed; the change happens in the disciples. In the transfiguration/resurrection, Christ does not become something that he was not previously (i.e. divine or deified)—it is the disciples who suddenly recognize this permanent feature of Christ. It is their perception of Christ that changes, not any ontology of Christ.

This is not a novel observation. Vladimir Lossky maintains that "a change was produced in the consciousness of the apostles", rather than Christ (Lossky, 1975, p. 61), as does George Maloney, who describes what took place in Peter, James, and John as a "radical change" (Maloney, 1968, p. 246). The transfiguration/resurrection, then, does not produce a change in Christ—Christ was always divine. Indeed, this book has taken pains to emphasize that no change can happen in God and that a kenotic incarnation does not deprive Christ of his divinity; God, being non-temporal, cannot change and so the incarnation is not Christ literally becoming "not-God" only for him to become God again after the resurrection. Rather, it is the disciples who realize that Christ is divine and are thus able to see Christ as he truly is (see Farrow, 1999, p. 181). In a similar vein, Küng writes:

> Is not the risen the same as the earthly Jesus? Must we not then say of the earthly what we say of the risen Jesus? Is not the earthly Jesus already the Son of God, even though his sovereignty is still hidden? (Küng, 1978, p. 390)

The resurrection does not change Christ; the resurrection changes the disciples. Christ's resurrection is no longer a raising from the dead in the sense of being a future objective transformation; Christ's resurrection was simply the disciples recognizing what he had "always" been. The resurrection is not God "doing something new" (indeed, God *can't* do anything "new"); the resurrection is the disciples' recognition of the single divine activity in Christ. Christ's resurrection is simply the subjective realization of the disciples that Christ is the Son of God.

In the transfiguration, which becomes a lens through which resurrection doctrine is viewed, no objective change happened in Christ, but a subjective change happened in the disciples. Regardless of whether Jesus actually rose from the dead or not, the locus of the resurrection is the change it brings in the *subject*. The resurrection from the dead of Jesus Christ, then, is not about Jesus at all: it is about the believers and their relationship with God. Jesus is certainly the object of resurrection faith, but, importantly, he is not the subject.

In a wider scope, the realization that Christ is the Son of God (i.e. "Jesus is risen") is also the realization that all creatures already imitate and participate in him; the recognition of Christ's divinity has universal implications. Such a view can be strengthened by the famous story about Seraphim of Sarov. In this, Motovilov complains of not being able to look at his master, Seraphim, because of Seraphim's transfiguration. Seraphim is shining with such a brilliant light that his disciple, Motovilov, cannot look directly at him. However, Seraphim replies that the reason he is shining with brilliant light is only because Motovilov himself has become transfigured. It is only because Motovilov was himself deified that he could perceive that deification of Seraphim of Sarov (Jakim, 2007, p. 255). Thomas Merton records a similar experience, complaining that "there is no way of telling people that they are walking around shining like the sun" (Merton, 1965, p. 157). It is easy to see the link with the transfiguration and, therefore, with the resurrection: Seraphim's brilliant shining is due to the subjective change, or deification, of the disciple Motovilov.[35] Thus, the only reason that the disciples could perceive the transfiguration of Christ is because they had become deified themselves.

Yet perhaps more important than this move to the subjective, the story of the transfiguration moves the resurrection away from the survival of death and towards relationship with God. This move away from survival of death and prolongation of life—more marked in the story of Seraphim of Sarov—also provokes the conclusion that resurrection is possible in "this life". Thus, for Ratzinger, the resurrection is the bond the individual has with Jesus *now* (Ratzinger, 1988, p. 117) and so:

> The true frontier between life and death does not lie in biological dying, but in the distinction between being with the one who is life and the isolation which refuses such "being-with". (Ratzinger, 1988, p. 207)

The resurrection, then, is nothing but relationship with God. Life is not a means to some other end; "what more could one want . . . than to live one's life in right relationship to God" (Scuka, 1989, p. 88). In this way, resurrection is synonymous with deification.

b. Resurrection as deification

On this view, there is nothing significant that separates resurrection from deification, which was described in Chapter 6 as participation and imitation. This means that resurrection is not a remedy for death and the prolongation of life but is the subjective threshold that turns participation in/imitation of Christ as creation into participation in/imitation of Christ as deification.

In Chapter 6 it was claimed that, due to the single nature of divine activity, creation and deification are not two events but one single event—it is only the creature, from the perspective of time, that can see two events. More properly, deification is not a separate act from creation but the subjective realization that the creature already participates in the divine life by virtue of being created; deification is not an objective, ontological transformation but the recognition that "this life" is a participation in the divine life. Resurrection, then, is the threshold between creation and deification; resurrection is the threshold between ignorance and recognition.

This connection between resurrection and deification is more a feature of Eastern than Western theology. Vladimir Lossky, for example, defines the "deified being" as someone who has "the complete consciousness of God" and "the fullness of grace appropriate to the age to come" (Lossky, 1975, p. 200). Nicholas Gendle, commenting on Gregory Palamas' *The Triads*, describes the deified person as someone who is "anticipating the resurrection-glory of the age to come" (Gendle, 1983, p. 138, n. 11). More importantly, however, this deification is possible now, in "this" life. Georgios Mantzaridis, for example, writes that deification is not a future reality "but a living reality in present existence" (Mantzaridis, 1984, p. 55).

However, again, this shifting of resurrection from a future reality to a present one is not exclusively an Eastern theology, but is present in such theologians as Moltmann and Torrance. Moltmann writes:

> Resurrection is not a consoling opium, soothing us with the promise of a better world in the hereafter. It is the energy for a rebirth of this life. The hope doesn't point to another world. It is focused on the redemption of this one. In the Spirit, resurrection is not merely expected. It is already experienced. Resurrection happens every day. (Moltmann, 1996, p. 81)

The rejection of resurrection as a future raising from the dead into perfection, then, is not only a result of the influence of neo-Darwinism: it is also rejected from a theological point of view. When taken in conjunction with the eternal, non-temporal divine nature and the contradiction of the resurrection narratives, it is not clear that the idea of resurrection as the future raising of an incorruptible body should even be understood as the "traditional" understanding of the resurrection.

More pertinently, this shifting of the resurrection from future reality to present experience could be seen as a biblical theme. Alister McGrath writes that the resurrection is both a "not yet" and an "already present", pointing to Paul's discussion of the Eucharistic celebration in 1 Corinthians 11:26 as being especially demonstrative of this tension (McGrath, 1987, p. 108 and p. 113). Jürgen Moltmann also points to the Bible to support his doctrine of the resurrection that "we already experience the power of the resurrection now in love" (Moltmann, 1996, p. 85), quoting John's first letter in which he writes that "we know that we have passed from death to life because we love one another. Whoever does not love abides in death" (1 John 3:14).

The idea of the resurrection as a physical, objective, future raising from the dead is not a necessary interpretation of resurrection. The Bible also recognizes that the creature's relationship with God is not rigidly correlative to the perfection of the body. This biblical tension between the "already present" and the "not yet" is important as a way to characterize the relationship between God and the creature.

Moreover, this tension is an infinite one, as has already been noted in relation to Gregory of Nyssa's doctrine of *epekstasis*. If the resurrection is "fullness of grace" and closeness to God, then there is always room for more grace and closeness to God. The resurrection cannot be an objective proximity to God, as this would make God susceptible to temporal dimensions, as has already been rejected. The threshold between creation and deification—which is resurrection—is therefore continually renewed and always, infinitely possible to a greater degree. The resurrection is always now but also always possible to a much greater degree.

In this way, therefore, the relationship that exists between the creature and God that forms the content of the doctrine of the resurrection is nothing but the same relationship that exists between the participated/imitated and the participator/imitator. Chapter 6, "Participation, Imitation, and Neo-Darwinism", has already expounded how deification can be understood along these lines and, now, this must apply equally to resurrection. The resurrected/deified is the creature that realizes it already imitates/participates in the life of God as the very ground of its being. As Henri de Lubac writes: "salvation consists in a personal ratification of [the creature's] original 'belonging' to Christ" (de Lubac, 1947, p. 39).

Resurrection and suffering

This understanding of the resurrection as a reinterpretation of the role of participation and imitation, as the subjective realization that the creature has always imitated God and participated in the divine life, as the subjective transformation of the creature's life rather than the objective transformation of its body, may very well fit with the neo-Darwinian context from which this book has argued. However, it does not fully answer the question of how the resurrection can deal with the inevitable suffering with which the previous chapter on the cross concluded. As Moltmann was quoted as saying, the resurrection is not a "consoling opium" (Moltmann, 1996, p. 81) that soothes the suffering of this life by promising an end to suffering in the next life.

As this chapter has consistently claimed, the suffering that comes with being open to failure and mistake is a permanent feature of the creaturely condition; suffering cannot be stopped, nor should the creature expect it to be. This inevitably leaves open the question of suffering with which this chapter started: what should be made of the suffering that blights and afflicts the lives of so many creatures?

Resurrection is not the transformation of the body to an incorruptible one that is incapable of suffering; rather, it is the transformation of the perception of, or outlook on, "this life". Suffering is not punishment for disobedience; it is the image in which creatures were created. The resurrection is not the objective ceasing of suffering, but the recognition and realization that the suffering brings the creature closer to God. To put it differently: the resurrection is not the reversal of the cross, but its validation and confirmation. In this way, Küng writes:

> Easter does not neutralize the cross but confirms it. The resurrection message therefore does not call for the adoration of a heavenly cult god who has left the cross behind him. It calls for imitation: to commit oneself in believing trust to this Jesus, to his message, and to shape one's own life in accordance with the standard of the crucified. The resurrection, that is, reveals the very thing that was not to be expected: that this crucified Jesus, despite everything, *was right*. (Küng, 1978, p. 382)

In a sense, then, the story of "doubting" Thomas is an important one. It is the *crucified* Jesus who is resurrected, complete with pierced hands and side. This means then, that the Jesus who is to be recognized as the form of the universe, the image all creatures imitate, is the *crucified* Jesus; it is the crucified Christ who creates, and it is the crucified Christ who is to be imitated. In fact, Torrance comes very close to what this book has argued, saying:

> In the resurrection it becomes revealed that Jesus Christ is none other than that almighty creator–Word of God, but with that disclosure the crucifixion is quite transformed in our understanding, as the way that God's creative activity has taken in the restoration of creation. (Torrance, 1976, p. 59)

The cross is a creative event, descriptive of and identical to the "original" act of creation in the incarnation. That act of creation defines the passibility and corruptibility of the creature and this inevitably leads to suffering, which is so characteristic of the creature's life (and also characteristic of neo-Darwinism). This ontological condition of suffering is also descriptive of the participation and imitation that characterizes the creature's relationship with God. The creature imitates God as the very ground of its being, but this imitation means that such imaging of God is never perfect and thus always open to failure and mistake.

Yet this suffering, by being the very image in which creation happens, is also the image that creatures should imitate in order to bring them closer to God. There is no objective difference between the suffering that characterizes the creature's life and the suffering that characterizes the creature's deification; it is only the subjective realization that the creature already participates in God that enables it to see the suffering of the world irradiating with blinding divine light.

Bibliography

Aghiorgoussis, M., 1992, "Orthodox Soteriology", in Meyendorff, J. and Tobias, R. (eds), *Salvation in Christ: A Lutheran-Orthodox Dialogue*, Minneapolis, MN: Augsburg Fortress Press.

Agourides, S., 1964, "The Social Character of Orthodoxy", in Philippou, A. (ed.), *The Orthodox Ethos: Studies in Orthodoxy Vol. 1*, Oxford: Holywell Press.

Aletti, J., 2002, "Romans 8: The Incarnation and its Redemptive Impact", in Davis, S., Kendall, D. and O'Collins, G. (eds), *The Incarnation*, Oxford: Oxford University Press.

Allchin, A., 1988, *Participation in God*, London: Darton, Longman & Todd.

Anderson, R., 1986, *Theology, Death & Dying*, Oxford: Basil Blackwell.

Anderson, R., 1998, "On Being Human: The Spiritual Saga of a Creaturely Soul", in Brown, W., Murphy, N. and Malony, H. N. (eds), *Whatever Happened to the Soul: Scientific and Theological Portraits of Human Nature*, Minneapolis, MN: Augsburg Fortress Press.

Anon., 1961, *The Cloud of Unknowing and Other Works*, Harmondsworth: Penguin Books.

Anselm of Canterbury, 1998, *Anselm of Canterbury: The Major Works*, Oxford: Oxford University Press.

Aquinas, T., 1948, *Summa Theologica Vol. 4*, New York: Benziger Brothers.

Aquinas, T., 1998, *Selected Writings*, London: Penguin Classics.

Athanasius, 1954, *Christology of the Later Fathers*, London: Westminster John Knox Press.

Augustine of Hippo, 1991, *Confessions*, Oxford: Oxford University Press.

Augustine of Hippo, 1991a, *The Trinity*, Hyde Park, NY: New City Press.

Aulén, G., 1931, *Christus Victor*, London: SPCK.

Ayala, F., 1998, "Human Nature: One Evolutionist's View", in Brown, W., Murphy, N. and Malony, H. N. (eds), *Whatever Happened to the Soul: Scientific and Theological Portraits of Human Nature*, Minneapolis, MN: Augsburg Fortress Press.

Ayres, L., 2004, *Nicaea and its Legacy*, Oxford: Oxford University Press.

Barbour, I., 1971, *Issues in Science and Religion*, New York, NY: Harper Torchbook.

Barbour, I., 2001, "God's Power: A Process View", in Polkinghorne, J. (ed.), *The Work of Life: Creation as Kenosis*, Cambridge: William B. Eerdmans.

Barclay, J., 1996, "The Resurrection in Contemporary New Testament Scholarship", in D'Costa, G. (ed.), *Resurrection Reconsidered*, Oxford: Oneworld.

Barron, R., 2007, *The Priority of Christ: Towards A Postliberal Catholicism*, Grand Rapids, MI: Brazos Press.

Barron, R., 2015, *Exploring Catholic Theology*, Grand Rapids, MI: Baker Academic.

Barron, R., 2016, *Vibrant Paradoxes*, Skokie, IL: World On Fire.

Barton G., 1934, *Christ and Evolution: A Study of the Doctrine of Redemption in the Light of Modern Knowledge*, Philadelphia, PA: University of Pennsylvania Press.

Bathrellos, D., 2004, *The Byzantine Christ: Person, Nature, and Will in the Christology of St. Maximus the Confessor*, Oxford: Oxford University Press.

Behe, M., 1998, *Darwin's Black Box*, New York, NY: Simon & Schuster.

Bernard, G., 2011, *Living Consciousness: The Metaphysical Vision of Henri Bergson*, New York, NY: SUNY Press.

Berry, R. J., 1982, *Neo-Darwinism*, London: Edward Arnold Limited.

Bettenson, H. and Maunder, C. (eds), 1999, *Documents of the Christian Church*, Oxford: Oxford University Press.

Bianchi, E., 1998, *Praying the Word of God*, Kalamazoo, MI: Cistercian Publications.

Birdsell, J. and Wills, C., 2003, "The Evolutionary Origin and Maintenance of Sexual Reproduction: A Review of Contemporary Models", *Evolutionary Biology* Vol. 33

Birx, H., 1972, *Teilhard de Chardin's Philosophy of Evolution*, Springfield, IL: Charles C. Thomas.

Birx, H., 1991, *Interpreting Evolution: Darwin & Teilhard de Chardin*, New York, NY: Prometheus Books.

Boethius, 2000, *The Consolation of Philosophy*, Oxford: Oxford University Press.

Boff, L., 1997, *Cry of the Earth, Cry of the Poor*, Maryknoll, NY: Orbis Books.

Bonaventure, 1978, *Bonaventure: The Soul's Journey to God, The Tree of Life, The Life of St. Francis*, Mahwah, NJ: Paulist Press.

Brown, D., 1985, *The Divine Trinity*, London: Duckworth.

Brown, R., 1973, *The Virginal Conception and The Resurrection of the Body*, New York, NY: Paulist Press.

Brown, W., 1998, "Cognitive Contributions to Soul", in Brown, W., Murphy, N. and Malony, H. N. (eds), *Whatever Happened to the Soul: Scientific and Theological Portraits of Human Nature*, Minneapolis, MN: Augsburg Fortress Press.

Burggren, W., 2014, "Epigenetics as a Source of Variation in Comparative Animal Physiology—Or—Lamarck is Lookin' Pretty Good These Days", *Journal of Experimental Biology* 217 (5).

Burns, J., 1981, *Theological Anthropology*, Philadelphia, PA: Fortress Press.

Bynum, W., 2009, "Introduction", in Darwin, C., *On the Origin of Species*, London: Penguin.

Cabasilas, N., 1974, *The Life in Christ*, Crestwood, NY: St Vladimir's Seminary Press.

Canale, F., 2009, *Creation, Evolution, and Theology: An Introduction to the Scientific and Theological Methods*, Santa Rosa, Argentina: Universidad Adventista del Plata Editorial.

Carey, N., 2012, *The Epigenetics Revolution*, London: Icon Books.

Cary, P., 2000, *Augustine's Invention of the Inner Self: The Legacy of a Christian Platonist*, Oxford: Oxford University Press.

Castelli, E., 1991, *Imitating Paul: A Discourse of Power*, Louisville, KX: Westminster/John Knox Press.

Cavanaugh, W., 1999, "Beyond Secular Parodies", in Milbank, J., Pickstock, C. and Ward, G. (eds), *Radical Orthodoxy*, Abingdon: Routledge.

CCC, 1997, *Catechism of the Roman Catholic Church*, Vatican City: Libreria Editrice Vaticana.

Chia, R., 2011, "Salvation as Justification and Deification", *Scottish Journal of Theology* 64:2.

Chown, M., 2014, "The Big Bang", in Webb, J. (ed.), *Nothing*, London: Profile Books.

Clement of Alexandria, 1969, "Protrepticus", in Bettenson, H. (ed.), *The Early Christian Fathers*, Oxford: Oxford University Press.

Coakley, S., 2001, "Kenosis: Theological Meanings and Gender Connotations", in Polkinghorne, J. (ed.), *The Work of Life: Creation as Kenosis*, Cambridge: William B. Eerdmans.

Coakley, S., 2002, "What Does Chalcedon Solve and What Does it Not? Some Reflections on the Status and Meaning of the Chalcedonian 'Definition'", in Davis, S., Kendall, D. and O'Collins, G. (eds), *The Incarnation*, Oxford: Oxford University Press.

Cole-Turner, R., 1993, *The New Genesis: Theology and the Genetic Revolution*, Louisville, KX: Westminster/John Knox Press.

Collins, F., 2007, *The Language of God*, London: Pocket Books.

Collins, G., 2010, *Meeting Christ in His Mysteries*, Dublin: Columba Press.

Conee, E. and Sider, T., 2005, *Riddles of Existence*, Oxford: Clarendon Press.

Cooper, J., 1989, *Body, Soul, and Life Everlasting: Biblical Anthropology and the Monism–Dualism Debate*, Cambridge, MA: William B. Eerdmans.

Corcoran, K., 2006, *Rethinking Human Nature*, Grand Rapids, MI: Baker Academic.

Corte, N., 1960, *Pierre Teilhard de Chardin: His Life and Spirit*, London: Barrie & Rockcliff.

Cortez, M., 2015, "The Madness in our Method: Christology as the Necessary Starting Point For Theological Anthropology", in Farris, J. and Taliaferro, C. (eds), *The Ashgate Research Companion to Theological Anthropology*, Farnham: Ashgate.

Coulson, C., 1958, *Science and Christian Belief*, London: Fontana Books.
Cowell, S., 2006, "Newman and Teilhard: The Challenge of the East", in Deane-Drummond, C. (ed.), *Teilhard de Chardin on People and Planets*, London: Equinox.
Cox, B. and Forshaw, J., 2009, *Why Does E=MC2?*, Cambridge: Da Capo Press.
Craig, W., 2001, *Time and Eternity: Exploring God's Relationship to Time*, Wheaton, IL: Crossway.
Cross, R., 2002, *The Metaphysics of the Incarnation*, Oxford: Oxford University Press.
Crysdale, C. and Ormerod, N., 2013, *Creator God, Evolving World*, Minneapolis, MN: Fortress Press.
Cuénot, C., 1967, *Science and Faith in Teilhard de Chardin*, London: Granstone Press.
Cyril of Jerusalem, 1982, "Catecheses", in Bettenson, H. (ed.), *The Later Christian Fathers*, Oxford: Oxford University Press.
Darwin, C., 2004, *The Descent of Man*, London: Penguin.
Darwin, C., 2009, *On the Origin of Species*, London: Penguin.
Davies, P., 1989, "Introduction", in Heisenberg, W., *Physics and Philosophy*, London: Penguin.
Davies, P., 1993, *The Mind of God*, London: Penguin.
Davies, P., 1996, *About Time*, New York, NY: Touchstone.
Davies, P., 2013, "The Day Time Began", in Webb, J. (ed.), *Nothing*, London: Profile Books.
Davis, S., 1993, *Risen Indeed: Making Sense of the Resurrection*, Grand Rapids, MI: W. B. Eerdmans Publishing.
Davis, S., 2010, "Is Kenosis Orthodox?", in Evans, C. (ed.), *Exploring Kenotic Christology: The Self-Emptying of God*, Vancouver: Regent College Publishing.
Davis, S., 2011, "The Metaphysics of Kenosis", in Hill, J. and Marmodoro, A. (eds), *The Metaphysics of the Incarnation*, Oxford: Oxford University Press.
Dawkins, R., 1986, *The Blind Watchmaker*, London: Penguin.
Dawkins, R., 1999, *The Extended Phenotype*, Oxford: Oxford University Press.

Dawkins, R., 2004, *A Devil's Chaplain*, London: Phoenix.
Dawkins, R., 2006, *The Selfish Gene*, Oxford: Oxford University Press.
Dawkins, R., 2006a, *The God Delusion*, London: Black Swan.
De Lubac, H., 1947, *Catholicism: Christ and the Common Destiny of Man*, San Francisco, CA: Ignatius Press.
De Lubac, H., 1956, *The Splendour of the Church*, San Francisco, CA: Ignatius Press.
De Lubac, H., 1966, *Teilhard Explained*, New York, NY: Paulist Press.
De Lubac, H., 1967, *The Religion of Teilhard de Chardin*, London: Collins.
De Lubac, H., 1969, *Augustinianism and Modern Theology*, London: Geoffrey Chapman.
De Lubac, H., 1971, *The Eternal Feminine*, London: Collins.
De Lubac, H., 1980, *A Brief Catechesis on Nature and Grace*, San Francisco, CA: Ignatius Press.
Deane-Drummond, C., 2009, *Christ and Evolution*, Minneapolis, MN: Fortress Press and London: SCM Press.
Dearman, J., 2002, "Theophany, Anthropomorphism, and the Imago Dei: Some Observations about the Incarnation in the Light of the Old Testament", in Davis, S., Kendall, D. and O'Collins, G. (eds), *The Incarnation*, Oxford: Oxford University Press.
Delfgaauw, B., 1969, *Evolution: The Theory of Teilhard de Chardin*, London: Fontana Books.
Delio, I., 2013, *The Unbearable Wholeness of Being*, Maryknoll, NY: Orbis Books.
Dennett, D., 1995, *Consciousness Explained*, Boston, MA: Little, Brown and Company.
Dilley, F., 1983, "Does the 'God Who Acts' Really Act?", in Thomas, O. (ed.), *God's Activity in the World: The Contemporary Problem*, Chico, CA: Scholars Press.
Dobzhansky, T., 1973, "Biology and the Human Condition", in Browning, G., Alioto, J. and Farber, S. (eds), *Teilhard de Chardin: In Quest of the Perfection of Man*, Rutherford, NJ: Farleigh Dickinson University Press.
Dobzhansky, T., 1982, *Genetics and the Origin of Species*, New York, NY: Columbia University Press.

Dodson, E., 1984, *The Phenomenon of Man Revisited*, New York, NY: Columbia University Press.
Dupré, L., 2009, "Intelligent Design: Science or Faith?", in Caruana, L. (ed.), *Darwinism and Catholicism*, London: T&T Clark.
Edwards, D., 1999, *The God of Evolution*, Mahwah, NJ: Paulist Press.
Einstein, A., 1954, *Relativity: The Special and the General Theory*, London: Routledge.
Elliot, H., 2012, "Introduction", in Lamarck, J. B., 2012, *Zoological Philosophy*, Memphis, TN: General Books.
Ellverson, A., 1981, *The Dual Nature of Man: A Study in the Theological Anthropology of Gregory of Nazianzus*, Stockholm: University of Uppsala.
Evans, C., 2002, "The Self-Emptying of Love: Some Thoughts on Kenotic Christology", in Davis, S., Kendall, D. and O'Collins, G. (eds), *The Incarnation*, Oxford: Oxford University Press.
Faricy, R., 1968, "The Theology of Human Endeavor", in Kessler, M. and Brown, B. (eds), *Dimensions of the Future: The Spirituality of Teilhard de Chardin*, Washington DC: Corpus Books.
Faricy, R., 1981, *All Things in Christ*, London: Fount.
Faricy, R., 1981a, "Introduction", in Maroky, P. (ed.), *Convergence*, Kottayam, India: Oriental Institute of Religious Studies.
Farrow, D., 1999, *Ascension and Ecclesia: On the Significance of the Doctrine of the Ascension for Ecclesiology and Christian Cosmology*, Edinburgh: T&T Clark.
Fee, G., 2002, "St. Paul and the Incarnation: A Reassessment of the Data", in Davis, S., Kendall, D. and O'Collins, G. (eds), *The Incarnation*, Oxford: Oxford University Press.
Ferguson, J., 1980, "In Defence of Pelagius", *Theology* 83:692.
Fiddes, P., 2000, *Participating in God*, London: Darton, Longman & Todd.
Fiddes, P., 2001, "Creation Out of Love", in Polkinghorne, J. (ed.), *The Work of Life: Creation as Kenosis*, Grand Rapids, MI: William B. Eerdmans.
Finch, J., 2006, "Irenaeus on the Christological Basis of Human Divinization", in Finlan, S. and Kharlamov, V. (eds), *Theosis: Deification in Christian Theology*, Cambridge: James Clarke & Co.

Finch, J., 2006a, "Athanasius on the Deifying Work of the Redeemer", in Finlan, S. and Kharlamov, V. (eds), *Theosis: Deification in Christian Theology*, Cambridge: James Clarke & Co.

Finlan, S. and Kharlamov, V. (eds), 2006, *Theosis: Deification in Christian Theology*, Cambridge: James Clarke & Co.

Finlan, S., 2007, "Can We Speak of Theosis in Paul?", in Christensen, M. and Wittung, J. (eds), *Partakers of the Divine Nature: The History and Development of Deification in the Christian Traditions*, Grand Rapids, MI: Baker Academic.

Fisher, R., 1930, *The Genetical Theory of Natural Selection*, Oxford: Oxford University Press.

Flanagan, O., 1991, *The Science of the Mind*, Cambridge, MA: The MIT Press.

Florovsky, G., 1975, *Aspects of Church History*, Belmont, MA: Norland Publishing Company.

Florovsky, G., 1976, *Creation & Redemption*, Belmont, MA: Norland Publishing Company.

Foster, C., 2009, *The Selfless Gene*, London: Hodder & Stoughton.

Fox, M., 1988, *The Coming of The Cosmic Christ*, New York, NY: HarperCollins.

Garcia-Rivera, A., 2009, *The Garden of God: A Theological Cosmology*, Minneapolis, MA: Fortress Press.

Gendle, N. (trans.), 1983, *Gregory Palamas: The Triads*, Mahwah, NJ: Paulist Press.

Gilkey, L., 1983, "Cosmology, Ontology, and the Travail of Biblical Language", in Thomas, O. (ed.), *God's Activity in the World: The Contemporary Problem*, Chico, CA: Scholars Press.

Gorman, M., 2009, *Inhabiting the Cruciform God*, Grand Rapids, MI: William B. Eerdmans.

Gould, S., 1982, "Introduction", in Dobzhansky, T., *Genetics and the Origin of Species*, New York, NY: Columbia University Press.

Gould, S., 1994, "The Evolution of Life", *Scientific American* 271 (4).

Gould, S., 1997, *Life's Grandeur*, London: Jonathan Cape.

Gould, S., 1999, *Rocks of Ages*, London: Vintage.

Gould, S., 2002, *The Structure of Evolutionary Theory*, Cambridge, MA: Belknap Press.

Gregersen, N., 2008, "Special Divine Action and the Quilt of Laws: Why the Distinction between Special and General Divine Action Cannot Be Maintained", in Russell, R., Murphy, N. and Stoeger, W. (eds), *Scientific Perspectives on Divine Action: Twenty Years of Challenge and Progress*, Vatican City: Vatican Observatory Foundation.

Gregory of Nazianzus, 1894, *Nicene and Post-Nicene Fathers* vol. 7, New York, NY: The Christian Literature Company.

Gregory of Nazianzus, 1954, *Christology of the Later Fathers*, London: Westminster John Knox Press.

Gregory of Nyssa, 1977, "Oratio Catechetica", in Bettenson, H., *The Later Christian Fathers*, Oxford: Oxford University Press.

Gregory of Nyssa, 1978, *The Life of Moses*, Mahwah, NJ: Paulist Press.

Griffith-Jones, E., 1909, *The Ascent Through Christ*, London: Hodder and Stoughton.

Grumett, D., 2005, *Teilhard de Chardin: Theology, Humanity, and Cosmos*, Leuven: Peeters.

Guttman, B., Griffiths, A., Suzuki, D. and Cullis, T., 2002, *Genetics*, Oxford: Oneworld, Oxford.

Haffner, P., 1995, *Mystery of Creation*, Leominster: Gracewing.

Hall, L., Rae, M. and Holmes, S. (eds), 2010, *SCM Reader Christian Doctrine*, London: SCM Press.

Ham, K., 1987, *The Lie: Evolution*, Green Forest, AR: Master Books.

Harris, S., 2003, *Understanding the Bible*, Boston, MA: McGraw Hill.

Hart, R., 1997, "Fault/Fall in Human Nature/Nurture?", in Rouner, L. (ed.), *Is There A Human Nature?*, Notre Dame, IN: University of Notre Dame Press.

Haught, J., 2000, *God After Darwin*, Boulder, CO: Westview Press.

Haught, J., 2003, "Chaos, Complexity, and Theology", in Fabel, A. and St. John, D. (eds), *Teilhard in the 21st Century*, Maryknoll, NY: Orbis Books.

Haught, J., 2010, "Teilhard and the Question of Life's Suffering", in Duffy, K. (ed.), *Rediscovering Teilhard's Fire*, Philadelphia, PA: St Joseph's University Press.

Haught, J., 2010a, *Making Sense of Evolution*, Louisville, KY: Westminster John Knox Press.

Hawking, S., 1988, *A Brief History of Time*, London: Bantam Press.

Hawking, S., 2010, *The Grand Design*, London: Bantam Press.

Hefner, P., 1993, *The Human Factor*, Minneapolis, MN: Augsburg Fortress.

Heisenberg, W., 1958, *Physics and Philosophy*, London: Penguin.

Herman, L., Kuczaj, S. and Holder, M., 1993, "Responses to Anomalous Gestural Sequences by a Language-Trained Dolphin: Evidence for Processing of Semantic Relations and Syntactic Information", *Journal of Experimental Psychology: General* 122 (2), 184–194.

Hilary of Poitiers, 1982, "On the Trinity", in Bettenson, H. (ed.), *The Later Christian Fathers*, Oxford: Oxford University Press.

Hill, J., 2003, *The History of Christian Thought*, Oxford: Lion Publishing.

Hillemann, F., Bugnyar, T., Kotrschal, K. and Wascher, C. A. F., 2014, "Waiting for better, not for more: corvids respond to quality in two delay maintenance tasks", *Animal Behaviour* 90, 1–10.

Hinlicky, P., 1997, "Theological Anthropology: Toward Integrating Theosis and Justification by Faith", *Journal of Ecumenical Studies* 34:1.

Hoekema, A., 1994, *Created in God's Image*, Grand Rapids, MI: William B. Eerdmans.

Holmes, S., 2011, "A Simple Salvation? Soteriology and the Perfections of God", in Davidson, I. and Rae, M. (eds), *God of Salvation: Soteriology in Theological Perspective*, Farnham: Ashgate.

Hume, B., 2002, *Searching for God*, York: Ampleforth Abbey Press.

Huxley, J., 1942, *Evolution: The Modern Synthesis*, London: George Allen & Unwin.

Irenaeus, 1969, "Against Heresies", in Bettenson, H. (ed.), *The Early Christian Fathers*, Oxford: Oxford University Press.

Jakim, B., 2007, "Sergius Bulgakov: Russian Theosis", in Christensen, M. and Wittung, J. (eds), *Partakers of the Divine Nature: The History and Development of Deification in the Christian Traditions*, Grand Rapids, MI: Baker Academic.

John of Damascus, 2013, *An Exact Exposition of the Orthodox Faith*, New Delhi: Isha Books.

Jonas, H., 1996, *Mortality and Morality: A Search for God After Auschwitz*, Evanston, IL: Northwestern University Press.
Juntunen, S., 1998, "Luther and Metaphysics: What is the Structure of Being According to Luther?", in Braaten, C. and Jenson, R. (eds), *Union with Christ: The New Finnish Interpretation of Luther*, Grand Rapids, MA: William B. Eerdmans.
Kärkkäinen, V., 2004, *One with God: Salvation as Deification and Justification*, Collegeville, MN: Liturgical Press.
Keating, D., 2007, *Deification and Grace*, Naples, FL: Sapientia Press.
Kelly, J. N. D., 1965, *Early Christian Doctrines*, London: Adam and Charles Black.
Kelly, J., 2009, *The Ecumenical Councils of the Catholic Church*, Collegeville, MN: Liturgical Press.
Kenney, W., 1970, *A Path Through Teilhard's Phenomenon*, Dayton, OH: Pflaum Press.
Kerr, F., 1997, *Immortal Longings: Versions of Transcending Humanity*, Notre Dame, IN: University of Notre Dame Press.
Kirkpatrick, F., 2014, *The Mystery and Agency of God: Divine Being and Action in the World*, Minneapolis, MN: Fortress Press.
Klauder, F., 1971, *Aspects of the Thought of Teilhard de Chardin*, North Quincy, MA: The Christopher Publishing House.
Krauss, L., 2012, *A Universe from Nothing*, London: Simon & Schuster.
Kropf, R., 1980, *Teilhard, Scripture, and Revelation*, Rutherford, NJ: Associated University Presses.
Küng, H., 1978, *On Being a Christian*, London: Fount Paperbacks.
Lacey, A., 1989, *Bergson*, London: Routledge.
Lamarck, J. B., 2012, *Zoological Philosophy*, Memphis, TN: General Books.
Lane, D., 1975, *The Reality of Jesus*, London: Sheed & Ward.
Lane, D., 1996, *The Phenomenon of Teilhard: Prophet for a New Age*, Macon, GA: Mercer University Press.
Leftow, B., 2002, "A Timeless God Incarnate", in Davis, S., Kendall, D. and O'Collins, G. (eds), *The Incarnation*, Oxford: Oxford University Press.

Long, S., 2011, *Analogia Entis: On the Analogy of Being, Metaphysics, and the Act of Faith*, Notre Dame, IN: University of Notre Dame Press.

Lossky, V., 1957, *The Mystical Theology of the Eastern Church*, Cambridge: James Clark & Co.

Lossky, V., 1975, *In the Image and Likeness of God*, Crestwood, NY: St Vladimir's Seminary Press.

Lossky, V., 1978, *Orthodox Theology: An Introduction*, Crestwood, NY: St Vladimir's Press.

Loughlin, G., 1999, "God's Sex", in Milbank, J., Pickstock, C. and Ward, G. (eds), *Radical Orthodoxy*, London: Routledge.

Louth, A., 2007, "The Place of Theosis in Orthodox Theology", in Christensen, M. and Wittung, J. (eds), *Partakers of the Divine Nature: The History and Development of Deification in the Christian Traditions*, Grand Rapids, MI: Baker Academic.

Lyons, J., 1982, *The Cosmic Christ in Origen and Teilhard de Chardin*, Oxford: Oxford University Press.

Mahoney, J., 2011, *Christianity in Evolution*, Washington D.C.: Georgetown University Press.

Malherbe, A. and Ferguson, E., 1978, "Introduction", in Gregory of Nyssa, *The Life of Moses*, Mahwah, NJ: Paulist Press.

Maloney, G., 1968, *The Cosmic Christ*, New York, NY: Sheed and Ward.

Mangalath, D., 1981, "The Teilhardian Spirituality", in Maroky, P. (ed.), *Convergence*, Kottayam, India: Oriental Institute of Religious Studies.

Mannermaa, T., 1998, "Why is Luther so Fascinating? Modern Finnish Luther Research", in Braaten, C. and Jenson, R. (eds), *Union with Christ: The New Finnish Interpretation of Luther*, Grand Rapids, MI: William B. Eerdmans.

Mannermaa, T., 1998a, "Justification and Theosis in Lutheran—Orthodox Perspective", in Braaten, C. and Jenson, R. (eds), *Union with Christ: The New Finnish Interpretation of Luther*, Grand Rapids, MI: William B. Eerdmans.

Mannermaa, T., 2005, *Christ Present in Faith: Luther's View of Justification*, Minneapolis, MN: Fortress Press.

Mantzaridis, G., 1984, *The Deification of Man*, Crestwood, NY: St Vladimir's Seminary Press.

Marshall, R., 1994, *The Transfiguration of Jesus*, London: Darton, Longman and Todd.

Marxsen, W., 1970, *The Resurrection of Jesus of Nazareth*, London: SCM Press.

Mattox, M. and Roeber, A., 2012, *Changing Churches: An Orthodox, Catholic, and Lutheran Theological Conversation*, Grand Rapids, MI: William B. Eerdmans.

McCabe, H., 2016, "Eternity", in McCosker, P. and Turner, D. (eds), *The Cambridge Companion to the Summa Theologiae*, Cambridge: Cambridge University Press.

McCarty, D., 1976, *Teilhard de Chardin*, Waco, TX: Ward Books.

McCord Adams, M., 1999, *What Sort of Nature? Medieval Philosophy and the Systematics of Christology*, Milwaukee, WI: Marquette University Press.

McDaniel, M., 1992, "Salvation as Justification and Theosis", in Meyendorff, J. and Tobias, R. (eds), *Salvation in Christ: A Lutheran-Orthodox Dialogue*, Minneapolis, MN: Augsburg Fortress.

McGrath, A., 1987, *The Enigma of The Cross*, London: Hodder & Stoughton.

McGrath, A., 2005, *Dawkins' God*, Oxford: Blackwell Publishing.

McGrath, A. and Collicutt McGrath, J., 2007, *The Dawkins Delusion*, London: SPCK.

McIntosh, M., 1998, *Mystical Theology*, Oxford: Blackwell Publishers.

McLeod Bryan, G., 1961, *In His Likeness*, London: SPCK.

Merton, T., 1961, *New Seeds of Contemplation*, London: Burns & Oates.

Merton, T., 1965, *Conjectures of a Guilty Bystander*, New York, NY: Doubleday.

Meyendorff, J., 1964, *A Study of Gregory Palamas*, London: The Faith Press.

Meyendorff, J., 1978, "Preface & Foreword", in Gregory of Nyssa, *The Life of Moses*, Mahwah, NJ: Paulist Press.

Meyendorff, J., 1983, *Gregory Palamas: The Triads*, Mahwah, NJ: Paulist Press.

Miller, K., 1999, *Finding Darwin's God*, New York, NY: HarperCollins.

Mivart, S., 1871, *On the Genesis of Species*, New York, NY: D. Appleton and Company.

Moltmann, J., 1990, *The Way of Jesus Christ*, London: SCM Press.

Moltmann, J., 1996, "The Resurrection of Christ: Hope for the World", in D'Costa, G. (ed.), *Resurrection Reconsidered*, Oxford: Oneworld.

Moltmann, J., 2001, "God's Kenosis in the Creation and Consummation of the World", in Polkinghorne, J. (ed.), *The Work of Life: Creation as Kenosis*, Grand Rapids, MI: William B. Eerdmans.

Moltmann, J., 2007, *The Future of Creation*, Minneapolis, MN: Fortress Press.

Monod, J., 1972, *Chance and Necessity*, London: Collins.

Montagnes, B., 2004, *The Doctrine of Analogy of Being According to Thomas Aquinas*, Milwaukee, WI: Marquette University Press.

Mooney, C., 1966, *Teilhard de Chardin and The Mystery of Christ*, London: Collins.

Moritz, J., 2015, "Evolutionary Biology and Theological Anthropology", in Farris, J. and Taliaferro, C. (eds), *The Ashgate Research Companion to Theological Anthropology*, Farnham: Ashgate.

Morris, S., 2003, *Life's Solution*, Cambridge: Cambridge University Press.

Morris, S., 2010, "Evolution and the Inevitability of Intelligent Life", in Harrison, P. (ed.), *The Cambridge Companion to Science and Religion*, Cambridge: Cambridge University Press.

Morris, T., 1986, *The Logic of God Incarnate*, Ithaca, NY: Cornell University Press.

Murphy, G., 2013, *Models of Atonement: Speaking about Salvation in a Scientific World*, Minneapolis, MN: Lutheran University Press.

Murphy, N., 1998, "Human Nature: Historical, Scientific, and Religious Issues", in Brown, W., Murphy, N. and Malony, H. N. (eds), *Whatever Happened to the Soul: Scientific and Theological Portraits of Human Nature*, Minneapolis, MN: Augsburg Fortress Press.

Murphy, N., 1998a, "Nonreductive Physicalism: Philosophical Issues", in Brown, W., Murphy, N. and Malony, H. N. (eds), *Whatever Happened to the Soul: Scientific and Theological Portraits of Human Nature*, Minneapolis, MN: Augsburg Fortress Press.

Murphy, N., 2008, "Emergence, Downward Causation, and Divine Action", in Russell, R., Murphy, N. and Stoeger, W. (eds), *Scientific Perspectives on Divine Action: Twenty Years of Challenge and Progress*, Vatican City: Vatican Observatory Foundation.

Nelstrop, L., 2009, *Christian Mysticism: An Introduction to Contemporary Theoretical Approaches*, Farnham: Ashgate.

Norris, F., 1996, "Deification: Consensual & Cogent", *Scottish Journal of Theology* 49:4.

Norris, R., 1996, "Chalcedon Revisited: A Historical and Theological Reflection", in Nassif, B. (ed.), *New Perspectives on Historical Theology: Essays in Memory of John Meyendorff*, Grand Rapids, MI: William B. Eerdmans.

North, R., 1968, "Tradition in Spirituality", in Kessler, M. and Brown, B. (eds), *Dimensions of the Future: The Spirituality of Teilhard de Chardin*, Washington D.C.: Corpus Books.

Numbers, R., 2010, "Scientific Creationism and Intelligent Design", in Harrison, P. (ed.), *The Cambridge Companion to Science and Religion*, Cambridge: Cambridge University Press.

O'Collins, G., 1983, *Interpreting Jesus*, London: Mowbray.

O'Collins, G., 1988, *Interpreting the Resurrection: Examining the Major Problems in the Stories of Jesus' Resurrection*, Mahwah, NJ: Paulist Press.

O'Collins, G., 1995, *Christology: A Biblical, Historical, and Systematic Study of Jesus*, Oxford: Oxford University Press.

O'Collins, G., 2002, "The Incarnation: The Critical Issues", in Davis, S., Kendall, D. and O'Collins, G. (eds), *The Incarnation*, Oxford: Oxford University Press.

O'Connell, R., 1994, *Soundings in St. Augustine's Imagination*, New York, NY: Fordham University Press.

O'Grady, J., 1985, *Heresy*, Shaftesbury: Element Books.

O'Leary, D., 2007, *Roman Catholicism and Modern Science*, New York, NY: Continuum.

O'Rourke, F., 1992, *Pseudo-Dionysius and the Metaphysics of Aquinas*, Notre Dame, IN: University of Notre Dame Press.

Oakes, E., 2016, *A Theology of Grace in Six Controversies*, Grand Rapids, MI: William B. Eerdmans.

Ogden, S., 1983, "What Sense Does It Make to Say, 'God Acts in History'?", in Thomas, O. (ed.), *God's Activity in the World: The Contemporary Problem*, Chico, CA: Scholars Press.

Origen, 1969, "Commentary on John", in Bettenson, H. (ed.), *The Early Christian Fathers*, Oxford: Oxford University Press.

Osborn, E., 2001, *Irenaeus of Lyon*, Cambridge: Cambridge University Press.

Palamas, G., 1983, *Gregory Palamas: The Triads*, Mahwah, NJ: Paulist Press.

Paley, W., 2006, *Natural Theology*, Oxford: Oxford University Press.

Palmer, F., 1949, *The Cross in History and Experience*, London: Church Book Room Press Ltd.

Parekh, B., 1997, "Is There a Human Nature?", in Rouner, L. (ed.), *Is There A Human Nature?*, Notre Dame, IN: University of Notre Dame Press.

Passmore, J., 1970, *The Perfectibility of Man*, London: Duckworth.

Peacocke, A., 1993, *Theology for a Scientific Age*, London: SCM Press.

Peacocke, A., 2001, *Paths from Science Towards God*, Oxford: Oneworld.

Peacocke, A., 2001a, "The Cost of New Life", in Polkinghorne, J. (ed.), *The Work of Life: Creation as Kenosis*, Grand Rapids, MI: William B. Eerdmans.

Peacocke, A., 2008, "Some Reflections on 'Scientific Perfections On Divine Action'", in Russell, R., Murphy, N. and Stoeger, W. (eds), *Scientific Perspectives on Divine Action: Twenty Years of Challenge and Progress*, Vatican City: Vatican Observatory Foundation.

Pelikan, J., 1971, *Historical Theology: Continuity and Change in Christian Doctrine*, London: Hutchinson & Co.

Pelikan, J., 1977, *The Christian Tradition: The Spirit of Eastern Christendom (600–1700)*, Chicago: University of Chicago Press.

Peters, T. and Hewlett, M., 2003, *Evolution from Creation to New Creation*, Nashville, TN: Abingdon Press.

Polkinghorne, J., 1988, *Science and Creation*, London: SPCK.

Polkinghorne, J., 1989, *Science and Providence*, London: SPCK.

Polkinghorne, J., 2001, "Kenotic Creation and Divine Action", in Polkinghorne, J. (ed.), *The Work of Life: Creation as Kenosis*, Grand Rapids, MI: William B. Eerdmans.
Pollard, W., 1958, *Chance and Providence*, London: Faber & Faber.
Powell, S., 2003, *Participating in God*, Minneapolis, MN: Augsburg Fortress Press.
Prat, F., 1945, *The Theology of St. Paul: Vol.1*, London: Burns, Oates & Washbourne.
Prat, F., 1945a, *The Theology of St. Paul: Vol. 2*, London: Burns, Oates & Washbourne.
Prior, H., Schwarz, A. and Güntürkün, O., 2008, "Mirror-Induced Behaviour in the Magpie (Pica pica): Evidence of Self-Recognition", *PLoS Biology* 6 (8): e202.
Pseudo-Dionysius, 1987, *The Complete Works*, Mahwah, NJ: Paulist Press.
Quick, O., 1938, *Doctrines of The Creed*, London: Nisbet & Co.
Radcliffe, T., 2005, *What is the Point of Being a Christian*, London: Burns & Oates.
Rahner, K., 1963, *Theological Investigations—Vol.2: Man in the Church*, London: Darton, Longman & Todd.
Rahner, K., 1966, *Theological Investigations—Vol.4: More Recent Writings*, London: Darton, Longman & Todd.
Rahner, K., 1966a, *Theological Investigations—Vol.5: Later Writings*, London: Darton, Longman & Todd.
Ratzinger, J., 1988, *Eschatology: Death and Eternal Life*, Washington D. C.: The Catholic University of America Press.
Riches, A., 2016, *Ecce Homo*, Grand Rapids, MI: William B. Eerdmans.
Riordan, W., 2008, *Divine Light*, San Francisco, CA: Ignatius Press.
Robinson, J., 1973, *The Human Face of God*, London: SCM Press.
Robinson, R., 1926, *The Christian Doctrine of Man*, Edinburgh: T&T Clark.
Rolston III, H., 2001, "Kenosis and Nature", in Polkinghorne, J. (ed.), *The Work of Life: Creation as Kenosis*, Grand Rapids, MI: William B. Eerdmans.
Rolston III, H., 2006, *Science and Religion*, Philadelphia, PA: Templeton Foundation Press.

Rovelli, C., 2015, *Seven Brief Lessons on Physics*, London: Penguin.
Russell, N., 2004, *The Doctrine of Deification in the Greek Patristic Tradition*, Oxford: Oxford University Press.
Russell, R., 2008, "Challenge and Progress in 'Theology and Science': An Overview of The VO/CTNS Series", in Russell, R., Murphy, N. and Stoeger, W. (eds), *Scientific Perspectives on Divine Action: Twenty Years of Challenge and Progress*, Vatican City: Vatican Observatory Foundation.
Saarinen, R., 1997, *Faith and Holiness: Lutheran–Orthodox Dialogue 1959–1994*, Göttingen: Vandenhoeck & Ruprecht.
Saunders, N., 2002, *Divine Action & Modern Science*, Cambridge: Cambridge University Press.
Savary, L., 2010, *The New Spiritual Exercises*, Mahwah, NJ: Paulist Press.
Schmidt, T., 2002, *A Scandalous Beauty*, Grand Rapids, MI: Brazos Press.
Scuka, R., 1989, "Resurrection: Critical Reflections on a Doctrine in Search of Meaning", *Modern Theology* 6 (1).
Seife, C., 2000, *Zero: The Biography of a Dangerous Idea*, London: Souvenir Press.
Sherrard, P., 1964, "The Sacrament", in Philippou, A. (ed.), *The Orthodox Ethos: Studies in Orthodoxy Vol.1*, Oxford: Holywell Press.
Shortt, R., 2016, *God Is No Thing*, London: Hurst & Company.
Shuster, M., 2002, "The Incarnation in Selected Christmas Sermons", in Davis, S., Kendall, D. and O'Collins, G. (eds), *The Incarnation*, Oxford: Oxford University Press.
Skow, B., 2015, *Objective Becoming*, Oxford: Oxford University Press.
Smith, W., 1988, *Teilhardism and the New Religion*, Rockford, IL: Tan Books and Publishers Inc.
Southgate, C., 2008, *The Groaning of Creation*, Louisville, KY: Westminster John Knox Press.
Stoeger, W., 2008, "Conceiving Divine Action in a Dynamic Universe", in Russell, R., Murphy, N. and Stoeger, W. (eds), *Scientific Perspectives on Divine Action: Twenty Years of Challenge and Progress*, Vatican City: Vatican Observatory Foundation.

Stoeger, W., 2010, "God, Physics, and the Big Bang", in Harrison, P. (ed.), *The Cambridge Companion to Science and Religion*, Cambridge: Cambridge University Press.
Stott, J., 1986, *The Cross of Christ*, Leicester: InterVarsity Press.
Tanner, K., 2011, "Creation and Salvation in The Image of an Incomprehensible God", in Davidson, I. and Rae, M. (eds), *God of Salvation: Soteriology in Theological Perspective*, Farnham: Ashgate.
Teilhard de Chardin, P., 1959, *The Phenomenon of Man*, New York: Harper & Row.
Teilhard de Chardin, P., 1968, *The Divine Milieu*, New York: Harper & Row.
Teilhard de Chardin, P., 1968a, *Writings in Time of War*, London: Collins.
Teilhard de Chardin, P., 1969, *Human Energy*, London: Collins.
Teilhard de Chardin, P., 1978, *The Heart of the Matter*, London: Collins.
Teilhard de Chardin, P., 1978a, *The Activation of Energy*, London: Harvest.
Teilhard de Chardin, P., 2004, *The Future of Man*, New York: Doubleday.
Tertullian, 1969, "Selections", in Bettenson, H. (ed.), *The Early Christian Fathers*, Oxford: Oxford University Press.
TeSelle, E., 1970, *Augustine the Theologian*, London: Burns and Oates.
Tinsley, E., 1960, *The Imitation of God in Christ*, London: SCM Press.
Torrance, T., 1976, *Space, Time, and Resurrection*, Edinburgh: The Handsel Press.
Torrance, T., 1978, *Space, Time, and Incarnation*, Oxford: Oxford University Press.
Torrance, T., 2001, *The Ground and Grammar of Theology*, Edinburgh: T&T Clark.
Tracy, T., 2008, "Special Divine Action and the Laws of Nature", in Russell, R., Murphy, N. and Stoeger, W (eds), *Scientific Perspectives on Divine Action: Twenty Years of Challenge and Progress*, Vatican City: Vatican Observatory Foundation.
Turner, D., 1998, *The Darkness of God: Negativity in Christian Mysticism*, Cambridge: Cambridge University Press.

Van Driel, E., 2006, "The Logic of Assumption", in Evans, C. S. (ed.), *Exploring Kenotic Christology*, Vancouver: Regent College Publishing.

Verschuuren, G., 2016, *Aquinas and Modern Science*, Kettering, OH: Angelico Press.

Vogel, L., 1996, "Introduction", in Jonas, H., *Mortality and Morality: A Search for God After Auschwitz*, Evanston, IL: Northwestern University Press.

Volz, C., 1992, "Human Participation in the Divine–Human Dialogue", in Meyendorff, J. and Tobias, R. (eds), *Salvation in Christ: A Lutheran–Orthodox Dialogue*, Minneapolis, MN: Augsburg Fortress.

Vonnegut, K., 2000, *Slaughterhouse 5*, London: Vintage.

Ward, K., 2001, "Cosmos and Kenosis", in Polkinghorne, J. (ed.), *The Work of Life: Creation as Kenosis*, Grand Rapids, MI: William B. Eerdmans.

Ward, K., 2006, *Pascal's Fire*, Oxford: Oneworld.

Weber, H., 1979, *The Cross: Tradition and Interpretation*, London: SPCK.

Wegter-McNelly, K., 2008, "Does God Need Room to Act?", in Russell, R., Murphy, N. and Stoeger, W. (eds), *Scientific Perspectives on Divine Action: Twenty Years of Challenge and Progress*, Vatican City: Vatican Observatory Foundation.

Weinert, F., 2009, *Copernicus, Darwin, and Freud*, Chichester: Wiley-Blackwell.

Weismann, A., 1893, *The Germ-Plasm: A Theory of Heredity*, London: Walter Scott Ltd.

Wiles, M. and Santer, M. (eds), 1975, *Documents in Early Christian Thought*, Cambridge: Cambridge University Press.

Wiles, M., 1972, "Does Christology Rest on a Mistake?", in Sykes, S. and Clayton, J. (eds), *Christ, Faith and History*, Cambridge: Cambridge University Press.

Wiles, M., 1974, *The Remaking of Christian Doctrine*, London: SCM Press.

Wiles, M., 1986, *God's Action in the World*, London: SCM Press.

Wilken, R., 2003, *The Spirit of Early Christian Thought*, London: Yale University Press.

Williams, A., 1999, *The Ground of Union: Deification in Aquinas and Palamas*, Oxford: Oxford University Press.

Williams, A., 2007, "Nestorianism: Is Jesus Christ one person or does he have a split identity, with his divine nature separate and divided from his human nature?", in Quash, B. and Ward, M. (eds), *Heresies and How to Avoid Them*, London: SPCK.

Williams, N., 1927, *The Ideas of The Fall and of Original Sin*, London: Longmans, Green & Co.

Williams, P., 2001, *Doing without Adam and Eve*, Minneapolis, MN: Fortress Press.

Williams, R., 1979, *The Wound of Knowledge*, London: Darton, Longman & Todd.

Williams, R., 1996, "Between the Cherubim: The Empty Tomb and the Empty Throne", in D'Costa, G. (ed.), *Resurrection Reconsidered*, Oxford: Oneworld.

Winter, M., 1995, *The Atonement*, London: Geoffrey Chapman.

Wood, S., 1998, *Spiritual Exegesis and the Church in the Theology of Henri de Lubac*, Edinburgh: T&T Clark.

Wright, N., 2002, "Jesus' Self-Understanding", in Davis, S., Kendall, D. and O'Collins, G. (eds), *The Incarnation*, Oxford: Oxford University Press.

Young, F., 1975, *Sacrifice and the Death of Christ*, London: SPCK.

Young, F., 1977, "A Cloud of Witnesses", in Hick, J. (ed.), *The Myth of God Incarnate*, London: SCM Press.

Young, F., 1992, *Can These Dry Bones Live?*, London: SCM Press.

Young, F., 2016, *Construing the Cross: Type, Sign, Symbol, Word, Action*, London: SPCK.

Young, P., 1973, "Law in the Age of Planetization", in Browning, G., Alioto, J. and Farber, S. (eds), *Teilhard de Chardin: In Quest of the Perfection of Man*, Rutherford, NJ: Farleigh Dickinson University Press.

Zagzubski, L., 2002, 'The Incarnation and Virtue Ethics', in Davis, S., Kendall, D. and O'Collins, G. (eds), *The Incarnation*, Oxford: Oxford University Press.

Zernov, N., 1964, "The Worship of the Orthodox and its Message", in Philippou, A. (ed.), *The Orthodox Ethos: Studies in Orthodoxy Vol. 1*, Oxford: Holywell Press.

Notes

[1] Teilhard makes the same point, but from the opposite perspective. He writes that "we still hear the word 'Darwinism' used as a synonym for 'evolution', as though what has happened in a mere half century goes for nothing" (Teilhard de Chardin, 1978a, p. 256). For Teilhard, treating the *fact* of evolution and the *mechanism* of Darwinism as synonyms is problematic because he believes there are better *mechanisms* (i.e. Lamarckism); however, this chapter argues that treating the *fact* of evolution and the *mechanism* of Darwinism as synonyms is problematic because it deceives people into assuming they are conforming to the specific and very well supported *mechanism* of neo-Darwinism simply by accepting the mutability of species.

[2] The famous example is that of the giraffe's neck evolving over a period of time due to its continual straining to reach higher leaves. However, Mivart provides a perfect example of the incorrectness of this theory by suggesting that circumcision would become unnecessary if Lamarck were correct (Mivart, 1871, p. 227). It is the gene that codes for the foreskin that is replicated, not the phenotype; therefore, subsequent generations still possess a foreskin despite the previous generation's removal of it. (It is interesting that the Darwinian rejection of Lamarckian-acquired characteristics could also be seen as a rejection of Augustinian original sin, whose theory of the sexual transmission of sin could be interpreted as an acquired characteristic.)

[3] There is a connection here with the existentialist doctrine of "existence before essence". For Existentialism, the individual exists first, before that existence is imbued with meaning; for Darwinism, the organ evolves first, before the body puts it to use.

[4] The presence of homologies—traits that are similar not because they perform a particular function but because they have been copied (see Dennett, 1995, p. 136)—is an excellent example of this. The human eye, for example, is not "perfect" as it contains a nerve that crosses the retina, causing a blind spot (a nerve that the cephalopod eye does not have, meaning it is "better" than

a human eye). If evolution was searching for perfection, it is assumed that it would have solved this "problem".

5   Paul Davies agrees that "the background of cosmic radiation [which consists of "extremely energetic particles from space that bombard the Earth incessantly"] induces mutations in biological organisms, and that helps to drive evolution, so in a sense we wouldn't be here without it. But too much would be a bad thing" (Davies, 1996, p. 55).

6   This also relegates the role of historical accident. For example, without the asteroid that caused the extinction of the dinosaurs, mammals would *probably* not have been afforded the opportunity to evolve. The evolution of mammals has as much to do with historical accident as it does with the accidental mutation of particular genes.

7   Dawkins argues that while genetic mutation is random, this does not mean that any mutation is possible. Going down one evolutionary path necessarily cuts off other evolutionary paths. Pigs, Dawkins argues, will *almost* certainly never be able to evolve wings, because the mutations required are unavailable to them (Dawkins, 1999, pp. 42ff.).

8   In this book the word "creature" is used as a neutral term that can apply to anything that is created. In this sense it includes everything from humanity, animals, mushrooms, rocks, "the sun, moon, and stars" (see Powell, 2003, p. 12), elementary particles (Polkinghorne, 1988, p. 72), and even time, space, matter, and energy. (Although, as it will be noted in the following chapter, time and space do not have an independent existence apart from matter/energy; there is not a "thing" called space in which matter is located, nor a "thing" called time in which events take place—both are "dimensions" of matter/energy.)

9   Due to the universal nature of Darwinism, this also includes non-living creatures: "there is no clear line between the living and the non-living" (Barbour, 1971, p. 6; see also Delfgaauw, 1969, p. 26).

10 The "Herring Gull—Black-backed Gull" ring (Miller, 1999, p. 47) provides a "spatial" example of the "temporal" link that Darwin describes.

11 In light of the impossibility of demarcating properly between living and non-living, this can be expanded to claim that the proton or the quark is equally our sister as the bacteriophage.

12 The non-uniqueness of humanity and unity of all creatures must include extra-terrestrial creatures. Francis Collins writes that "if God exists, and seeks

to have fellowship with sentient beings like ourselves, and can handle the challenge of interactivity with 6 billion of us currently on this planet and countless others who have gone before, it is not clear why it would be beyond his abilities to interact with similar creatures on a few other planets or, for that matter, a few million other planets" (Collins, 2007, p. 71).

13   N. P. Williams notes that this mistranslation is even quoted in Roman Catholic literature well into the twentieth century (Williams, 1927, p. 309), indicating just how far-reaching this problem is.

14   Although, importantly, this should not be understood as the narrow, one-dimensional "biblical literalism/fundamentalism" that characterizes modern approaches to biblical interpretation. The literal approach to biblical interpretation, especially for Augustine (see Wood, 1998, pp. 25ff.), was part of a wider, multi-dimensional approach to interpretation and exegesis that included allegorical and Christological approaches. Modern Protestantism, with its principle of "*sola scriptura*", is less in a position to accept this varied and multifaceted approach to biblical interpretation, seeing the literal interpretation as the only interpretation.

15   It is interesting that Robert Scuka also uses this same interpretation of the resurrection. He writes that "it is plausible to argue, however, that this is to get the historical relationship backwards; that it rather was the Christian's actual experience of new life in the Spirit and liberation from bondage that was the basis for the elaboration of the story about Jesus' resurrection" (Scuka, 1989, p. 78), which anticipates the conclusion to the chapter on resurrection in this book that resurrection is not a future raising of a perfect body but a present relationship with God.

16   This also means that time and space are not creatures in and of themselves; they do not have an independent existence. There is not a "thing" called space in which creatures live—as Einstein writes, "physical objects are not in space, but these objects are spatially extended. In this way the concept of empty space loses its meaning" (Einstein, 1954, p. vi)—nor is there a "thing" called time that exists apart from the individual creature that experiences it. Torrance, too, writes that "[time and space] are not receptacles apart from bodies or forces, but are functions of events in the universe and forms of their orderly sequence and structure" (Torrance, 1976, p. 130). Likewise, the two dimensions cannot be separated or treated apart from one another, as Skow writes: "there is no such thing as space 'by itself', something completely

separate from time. For now, places and times are made of the same things, points of spacetime, and any point of space and instant of time have a spacetime point in common" (Skow, 2015, p. 9).

17   Just as the term "creature" applies to anything that is created, so this book uses the term "created activity" to mean anything that is done by a creature—from quantum fluctuations and chemical reactions to deliberate human actions.

18   Incidentally, this can help to further elucidate what the neo-Darwinian synthesis claims about humanity's relationship with non-human creatures. Whereas God and humanity are equivocal, humanity and non-human creatures are univocal. Humanity is not ontologically distinct from non-human creatures; human intelligence, for example, is qualitatively the same as but quantitatively distinct from non-human intelligence, whereas creaturely and divine intelligence are so qualitatively distinct that God cannot be said to have intelligence.

19   Although, strictly speaking, there is not one identifiable "first moment of time": "time does not switch on abruptly, but emerges continuously from space—there is no specific first moment at which times starts, but neither does time extend backwards infinitely" (Davies, 2013, p. 51). The universe does not have a first moment, but it does have a temporal boundary.

20   This, of course, does not include the three persons of the Trinity, who are not parts in any sense of that word. All three persons are, in and of themselves, fully divine and wholly and completely united. This does not mean that the language of three (which obviously suggests complexity rather than simplicity) is not appropriate—it just means that it must always be metaphorical and analogical, bearing in mind that whatever similarities the created language points to in the divine, they are always trumped by a greater dissimilarity.

21   Clearly, the role of the Spirit is important here; however, this book will not consider this. Focusing on the uniqueness of the incarnation (and thus the non-materiality of the Spirit), this book will imply that the Spirit has no role to play in divine activity. This obviously creates problems for traditional theology, but this book will offer no solutions; it is enough to note that it is a problem with the theology that this book supports.

22   After all, it has already been argued that a sober re-interpretation of the divine nature means that such "traditional" attributes as omnipotence and omniscience must be rejected. These are also rejected on the basis of the rejection of the univocal relationship between God and creation—omnipotence

does not mean that God has an infinite version of the power that creatures have (i.e. a perfect version of the ability to influence the matrix of cause and effect).

23 The universe is not created by God, directly or immediately, so God's existence cannot be deduced, logically, by simply looking at the universe.

24 Presumably, Barron acknowledges that "even the simplest act of cognition" takes place through Christ, even if the individual is ignorant of Christ, which means that there are interesting comparisons between this idea and Karl Rahner's "anonymous Christians".

25 Of course, it has been argued that divine foreknowledge of such historical events would mean that God could have included his plan of salvation in the original intention for creation, knowing that sin would occur. However, the application of temporal language to God (i.e. *fore*knowledge), the implication that God's will is complex (i.e. the divine will is not simple), and the idea that God can "know" something (i.e. the postulation of a material brain to God due to the rejection of an immaterial consciousness not completely correlated to brain states and the inability of that material brain to gain information without the senses), all strongly reject the idea that God could have had foreknowledge of an historical Fall.

26 "Infinity" is quantitatively different (i.e. God has more of the *same* thing that creatures do), whereas "eternity" is qualitatively different (i.e. God has something totally *different*); this distinction between infinity and eternity is the same as that between univocal and equivocal. Likewise, infinity is endless, whereas eternity is nothing.

27 It is interesting to note here that if Dobzhansky is correct that "there is no single human nature; there are as many human natures as there are men" (Dobzhansky, 1973, p. 105), then Christ would not be relevant to anyone. It is essential that Christ shares in the nature of those to whom his incarnation is addressed otherwise the fruits of his incarnation are ineffective. If, on the other hand, there is one "nature" that characterizes not just what it means to be human, but what it means to be created, then all creatures are addressed in the incarnation. The fact that such a single nature is manifested as human does not mean that non-human creatures do not also share in that nature.

28 While there are no objective criteria by which creatures imitate Christ, and it would be impossible to speculate in what ways other creatures imitate Christ, the use of fractals and the "golden ratio" almost certainly opens up

an interesting avenue. In this way, Jesus becomes Leonardo da Vinci's famous "Vitruvian Man": the "template" or "model" which all creatures imitate. This is, perhaps, further supported by the fact that the Greek word "*logos*" can be translated as "ratio" just as much as "word" (Seife, 2000, p. 26); Christ is the "ratio" through which all creatures are created.

[29] Incidentally, the ontology of imitation does not just answer the question of why there is something rather than nothing, but also why this universe exists rather than any other. The only way to be created is to be open to failure and mistake, so the world we inhabit is not "the best possible world" or "the worst possible world" or a "mediocre world", it is the *only* possible world. This could be put differently: it is because the world is the way it is that explains why there is something rather than nothing; if the world were any different, it couldn't have been created. The two questions—"why is there something rather than nothing?" and "why is the universe this universe and not another?"—have the same answer: the incarnation and the ontology of imitation.

[30] It is important to remember that, as it has already been noted, time is a dimension in precisely the same way that space is and, in that way, the divine influence can be passed "backwards" in time just as much as "forwards" (see Conee and Sider, 2005, p. 57 and Craig, 2001, p. 196); as will be seen in the chapter on resurrection, the past and the future are just as existent as the present. The fact that the incarnation happens in the "middle" of time does not mean that it cannot influence the "first" moment of time.

[31] The fact that evolution is not a theory of creation, but ontology, also means that God does not *use* evolution as a way to create; this is not a theory of traditional deism whereby God sets up the universe with certain values and potentialities, that will become actual in the future (this applies temporal categories to God). Evolutionary change happens because of the ontology of imitation (i.e. imperfect replication) at a biological level, but this is accidental or incidental to God's act of creation. God's will/intention is just as simple as God's energy/action; God self-empties (i.e. the incarnation), which defines/grounds creatures (i.e. the ontology of imitation), the rest is accidental.

[32] The incarnation is an eternal event, and so does not happen "before" or "after" anything, but the birth of Christ, which happens in time, does happen before the crucifixion.

33   "Re-presentation" rather than "repetition"; the sacraments are not the same event *again*—so that there are many Calvaries—but actually the same event.
34   Although there is an important difference between the incarnation and the epiclesis in that the incarnation was the creation of the body and blood of Christ, whereas the epiclesis is the changing-into of the body and blood of Christ. The epiclesis, if it is an incarnation, cannot escape adoptionism.
35   It may be possible to reach the same conclusions regarding the stigmatics in the West. David Brown writes that "in the west the summit of their experience is commonly seen as the receiving of the stigmata, whereas in the east it is conceived of in terms of being bathed in light, the difference being due to the different doctrinal traditions, with the west emphasizing the crucifixion and the east the transfiguration" (Brown, 1985, p. 28). The context of this passage is one of contrast—the West understands experience of God differently to the East—but it may be possible to understand the differences as being arbitrary in the light of the similarities in terms of experience of God.

Index of Names

Aghiorgoussis, Maximos 64, 125
Agourides, Savvas 62
Anderson, Ray 58, 210
Anselm of Canterbury 60, 101, 135, 179, 183
Aquinas, Thomas 17, 58, 83–5, 86, 87, 89, 95, 96, 101, 114, 116–8, 124, 142, 148, 152, 153, 158, 171
Athanasius of Alexandria 105, 114, 125, 126
Augustine of Hippo 17, 57, 60–8, 69, 73, 76, 88–9, 95, 96, 101, 105, 124, 127, 134, 148, 149, 152, 157, 160, 178, 183, 188, 210, 258n14
Aulén, Gustav 177, 178, 179, 180, 183, 184, 187
Ayres, Lewis 127, 157

Barbour, Ian 94, 137, 147, 257n9
Barclay, John 208
Barron, Robert 85–6, 105, 117–8, 260n24
Barton, George 9
Behe, Michael 33–4
Berry, Robert 27, 34, 42
Bianchi, Enzo 106
Birdsell, John and Willis, Christopher 40
Birx, H. James 13, 21, 25, 85, 202
Boethius 86
Boff, Leonardo 18, 55
Brown, David 133–4, 212, 262n35
Brown, Raymond 209
Brown, Warren 57, 59
Burggren, Warren 38
Burns, J. Patout 65
Bynum, William 22

Cabasilas, Nicolas 115
Canale, Fernando 11
Carey, Nessa 38
Cary, Phillip 148
Castelli, Elizabeth 159, 161
Cavanaugh, William 162–3
Chia, Roland 63
Clement of Alexandria 108–9
Coakley, Sarah 135
Cole-Turner, Ronald 25, 29, 52, 71
Collins, Francis 28, 34, 67, 257–8n12
Collins, Gregory 152, 185, 192, 197
Conee, Earl and Sider, Theodore 90, 261n30
Cooper, John 58–9
Corte, Nicholas 43–4
Coulson, C. A 55
Cowell, Sion 60, 64
Cox, Brian and Forshaw, Jeff 82
Craig, William 78, 89, 91–2, 99, 261n30
Cross, Richard 110–1
Crysdale, Cynthia and Ormerod, Neil 10, 25–6, 42
Cuénot, Claude 68, 72
Cyril of Alexandria 110, 147, 155

Darwin, Charles 8, 13, 14, 21–2, 24–9, 35, 40, 42, 47, 49, 52–3, 76, 121, 140, 167–8, 202, 257n10
Davies, Paul 89, 138, 257n5, 259n19
Davis, Stephen 201
Dawkins, Richard 15, 16, 23, 25, 32–3, 35, 39–40, 42, 44, 52, 53–4, 76, 79–80, 82, 91, 140, 167–170, 172–3, 257n7
De Lubac, Henri 139, 151, 152, 216

INDEX OF NAMES

Deane-Drummond, Celia 10, 23, 52, 69, 131, 186, 190
Delfgaauw, Bernard 257n9
Delio, Ilia 17, 67, 72, 109
Dennett, Daniel 256n4
Dobzhansky, Theodosius 28, 29–30, 36, 40, 41, 169, 260n27
Dodson, Edward 52
Duns Scotus 85, 124
Dupré, Louis 34

Edwards, Denis 11
Einstein, Albert 17, 77–8, 99, 131, 204, 258n16
Elliot, Hugh 22
Ellverson, Anna-Stina 105
Evans, C. Stephen 134

Faricy, Robert 69
Farrow, Douglas 96, 213
Fee, Gordon 189
Ferguson, John 61
Fiddes, Paul 81, 131
Finch, Jeffrey 126, 147
Finlan, Stephen 94
Fisher, Ronald 29, 41, 52
Flanagan, Owen 81
Florovsky, Georges 118, 122, 124, 210
Foster, Charles 43, 60, 169

Garcia-Rivera, Alejandro 12
Gendle, Nicholas 125, 215
Gilkey, Langdon 85–6
Gorman, Michael 134, 196
Gould, Stephen Jay 13, 21–2, 35–6, 54, 71, 171
Gregersen, Niels 97
Gregory of Nazianzus 58, 87, 105, 126–7, 141, 181–2
Gregory of Nyssa 70, 125–6, 134, 150–1, 160, 165, 180, 216
Griffith-Jones, Ebenezer 10, 12
Guttman, Burton and Griffiths, A and Suzuki, David and Cullis, Tara 40, 42

Haffner, Paul 60, 97

Hall, Lindsey and Rae, Murray and Holmes, Steve 49, 58, 105
Harris, Stephen 63, 178
Hart, Ray 201
Haught, John 11–2, 24, 45, 131
Hawking, Stephen 84, 138
Hefner, Phillip 9–10, 50, 70, 75
Herman, Louis and Kuczaj, Stan and Holder, Mark 57
Hilary of Poitiers 127, 133–4
Hill, Jonathan 60
Hillemann, Friederike and Bugnyar, Thomas and Kotrschal, Kurt and Wascher, Claudia 57
Hinlicky, Paul 60
Hoekema, Anthony 49, 158, 163
Holmes, Stephen 93
Hume, Basil 99, 128, 137, 147, 151, 158, 164
Huxley, Julian 22, 37, 42

Irenaeus of Lyons 64–5, 105, 108–9, 123, 125, 147

Jakim, Boris 213
John of Damascus 111, 151
Jonas, Hans 44, 130–1
Juntunen, Sammeli 90

Keating, Daniel 148–150
Kelly, Joseph 66
Kelly, John N.D. 95, 115, 126
Kenney, W. Henry 52
Kerr, Fergus 73
Kirkpatrick, Frank 79–80, 83, 85, 87
Klauder, Francis 124
Krauss, Lawrence 131, 132, 138
Kropf, Richard 139
Küng, Hans 206, 208, 213, 217

Lacey, Alan 25
Lamarck, Jean-Baptiste 256n2
Lane, David 97,
Lane, Dermot 208
Leftow, Brian 140
Lossky, Vladimir 115, 146, 212, 215

Loughlin, Gerard 160
Louth, Andrew 123, 147

Mahoney, Jack 12, 90–1
Malherbe, Abraham and Ferguson, Everett 150
Maloney, George 7, 106, 126, 212
Mannermaa, Tuomo 154
Mantzaridis, Georgios 123, 125, 215
Marshall, Rob 212
Marxsen, Willi 211
Mattox, Mickey and Roeber, A. G. 60, 63
Maximus the Confessor 151
McCabe, Herbert 87
McCord Adams, Marilyn 107, 171, 183
McDaniel, M 151
McGrath, Alister 23, 32, 43, 82, 169, 176, 196, 198, 215–6
McIntosh, Mark 196
McLeod Bryan, G 166
Merton, Thomas 108, 111, 152–3, 213–4
Meyendorff, John 115, 132, 148, 150
Miller, Kenneth 34, 36, 257n10
Mivart, St. George 8, 9, 43, 170, 256n2
Moltmann, Jürgen 131, 186, 215–6, 217
Monod, Jacques 25, 30–2, 36, 41–2
Montagnes, Bernard 87, 117, 157, 160
Mooney, Christopher 139
Moritz, Joshua 54
Morris, Simon Conway 10, 36–7, 43, 54
Morris, Thomas 138
Murphy, George 187
Murphy, Nancey 56, 58, 81

Nelstrop, Louise 164, 172
Norris, F. 154
Norris, Richard 113
North, Robert 69
Numbers, Ronald 35

O'Collins, Gerald 107, 110, 112, 117, 125–6, 136, 160, 179, 188
O'Connell, Robert 105
O'Grady, Joan 61
O'Leary, Don 22
O'Rourke, Fran 163, 164–5

Oakes, Edward 72, 129
Ogden, Schubert 116
Origen of Alexandria 139, 147
Osborn, Eric 147

Palamas, Gregory 124–5, 215
Palmer, F. Noel 179, 188
Passmore, John 60, 87
Peacocke, Arthur 10, 22, 52, 76, 77, 131, 183, 187
Pelagius 60–1, 64, 73
Pelikan, Jaroslav 62, 64, 106, 115, 180
Peters, Ted and Hewlett, Martinez 12, 52
Polkinghorne, John 10–1, 81, 93, 100, 131, 257n8
Powell, Samuel 147, 152, 153, 155, 257n8
Prat, Ferdinand 190
Prior, Helmut and Schwarz, Ariane and Güntürkün, Onur 56
Pseudo-Dionysius 88, 98, 101, 151, 162, 164, 172, 174

Quick, Oliver 177, 188, 190, 210

Rahner, Karl 148, 260n24
Ratzinger, Joseph 205, 214
Riches, Aaron 111–2, 113–4
Riordan, William 98
Robinson, John 141
Rolston III, Holmes 11, 153
Rovelli, Carlo 203
Russell, Norman 147, 148, 149, 154–5, 156–7
Russell, Robert 77

Saarinen, Risto 154, 193
Saunders, Nicholas 42, 77, 79
Schmidt, Thomas 185
Scuka, Robert 98, 151, 152, 214, 258n15
Seife, Charles 261n28
Sherrard, Philip 194
Shortt, Rupert 71, 86, 108
Shuster, Marguerite 190
Skow, Bradford 258–9n16
Smith, Wolfgang 28, 43

Southgate, Christopher 12, 50
Stoeger, William 94–5
Stott, John 179, 180, 183, 190

Tanner, Kathryn 157
Teilhard de Chardin, Pierre 10, 51–2, 56, 68–9, 97, 124, 127, 137, 177, 186–7, 188, 193, 196, 256n1
Tertullian 62–3, 109, 178
TeSelle, Eugene 68
Torrance, Thomas 117, 134, 137, 215, 218, 258n16
Tracy, Thomas 61
Turner, Denys 86, 87, 112

Van Driel, Edwin 171
Verschuuren, Gerard 83–4, 99–100
Vogel, Lawrence 82
Volz, C 59
Vonnegut, Kurt 203–4

Ward, Keith 131

Weber, Hans-Ruedi 185
Wegtner-McNelly, Kirk 83
Weinert, Friedel 14–15, 22, 26, 35
Weismann, August 26
Wiles, Maurice 95, 97, 104, 106–7, 111
Wilken, Robert 50, 65, 123
William of Occam 85
Williams, A. N 98–9, 121, 125, 128, 146, 148, 151–2, 164
Williams, Norman P. 65, 70, 258n13
Williams, Patricia 26, 29, 33, 36, 42, 57, 63–4, 66–7, 70, 188
Williams, Rowan 197
Winter, Michael 59–60, 62–3, 121
Wood, Susan 157, 258n14
Wright, N. T. 107

Young, Frances 103–4, 142, 182, 187, 188, 194–5, 200

Zagzubski, Linda 165–6
Zernov, Nicolas 194

EU GPSR Authorized Representative:

LOGOS EUROPE, 9 rue Nicolas Poussin, 17000 La Rochelle, France

contact@logoseurope.eu